SpringerBriefs in Computer Science

More information about this series at http://www.springer.com/series/10028

Virinchi Srinivas · Pabitra Mitra

Link Prediction in Social Networks

Role of Power Law Distribution

Springer

Virinchi Srinivas
Department of Computer Science
University of Maryland
College Park, MD
USA

Pabitra Mitra
Department of Computer Science
  and Engineering
Indian Institute of Technology Kharagpur
Kharagpur, West Bengal
India

ISSN 2191-5768   ISSN 2191-5776   (electronic)
SpringerBriefs in Computer Science
ISBN 978-3-319-28921-2   ISBN 978-3-319-28922-9   (eBook)
DOI 10.1007/978-3-319-28922-9

Library of Congress Control Number: 2015960423

Printed on acid-free paper

This Springer imprint is published by SpringerNature
The registered company is Springer International Publishing AG Switzerland

Preface

Overview

Link prediction problem has been widely studied in the past. It is primarily applied in recommender systems. The problem deals with predicting new links that are likely to emerge in a network in the future, given the network at the current time. Two nodes tend to get connected if they are similar to each other. Hence, it is important to compute similarity between two nodes to assess the possibility of link formation between them. Some of the link prediction similarity measures such as *Common Neighbors* (*CN*), *Adamic Adar Index* (*AA*), *Resource Allocation Index* (*RA*), *Preferential Attachment Index* (*PA*), *Katz*, and *PropFlow* use the structure of the network like the neighborhood information and path information between a pair of nodes to compute similarity between two nodes. Higher similarity between two nodes implies higher possibility of link formation between the two nodes. Although power law degree distribution is an important notion in social networks, its role has not been explored adequately in the context of link prediction.

In this book, we propose link prediction similarity measures for social networks which exploit the degree distribution of the networks. In the context of link prediction in dense networks, we propose *Markov Inequality Degree Thresholding-based similarity measure* (*MIDT*), which only considers nodes whose degree is below a threshold for a possible link. The threshold value is determined using the Markov Inequality. Next, we present similarity measures based on cliques (CNC, AAC, RAC), which assigns extra weight between nodes sharing more number of cliques. Further, we propose a *locally adaptive* (*LA*) similarity measure, which assigns different weights to common nodes based on the degree distribution of the local neighborhood and the degree distribution of the network. The weight of the common node varies with the neighborhood under consideration. In the context of link prediction in sparse networks, we propose a novel *two-phase framework* that adds edges to the sparse graph to form a boost graph. We use the boost graph instead of the original network for link prediction.

Audience

This book is intended for graduate students and researchers working on the link prediction problem. Specifically, we present similarity measures in link prediction by exploiting the *power law degree distribution* of the network that social networks typically follow. The techniques could be applied in different areas such as computer science, engineering, etc. It is assumed that the reader has a basic knowledge of mathematics at the high-school level, as well as a certain background in computing and programming. Although the algorithms can be implemented in programming languages, such as C++, Java and Python, etc, C++ implementation renders superior performance in terms of computational speed.

Organization

This book is organized as follows:

1. *Literature and Background*: Chapter 1 presents the literature and state-of-the-art techniques in link prediction. Further, we also discuss the relevant background required for link prediction.
2. *Link Prediction in Dense Networks*: Chapters 2 and 3 present link prediction similarity measures based on power law degree distribution of the network in dense networks.

   (a) *Degree Thresholding*: Chapter 2 presents a node thresholding approach (MIDT) based on its degree. It assigns a smaller weight to the low-degree common nodes and ignores the contribution of the high-degree common nodes. Further, in Chap. 2, we also present a clique-based approach (CNC, AAC, RAC) to link prediction, which assigns heavier weights to node pairs that share common neighbors having more number of cliques.
   (b) *Locally Adaptive Approach*: Chapter 3 presents a Locally Adaptive (LA) Approach wherein the weights to the common nodes for similarity computation is based on the degree distribution of the local neighborhood and the degree distribution of the network. Further, the weights assigned to the common nodes vary with the local neighborhood under consideration.

3. *Link Prediction in Sparse Networks*: We present a framework for link prediction, a novel two-phase approach, to deal with sparse networks in Chap. 4.
4. *Applications of Link Prediction*: We present and discuss various applications of link prediction in Chap. 5.
5. *Conclusion*: We conclude in Chap. 6 and also present potential future directions for link prediction based on power law degree distribution of the networks.

Virinchi Srinivas
Pabitra Mitra

Contents

Acronyms

AA   Adamic Adar Index
AAC   Adamic Adar Index Using Cliques
AUC   Area Under Curve
CC   Clustering Coefficient
CN   Common Neighbor Index
CNC   Common Neighbor Index Using Cliques
KL   Kullback–Leibler Divergence
LA   Locally Adaptive Similarity Measure
MIDT Markov Inequality-based Degree Thresholding Similarity Measure
PA   Preferential Attachment Index
RA   Resource Allocation Index
RAC   Resource Allocation Index Using Cliques

Chapter 1
Introduction

Abstract Link prediction deals with predicting new links which are likely to emerge in network in the future, given the network at the current time. It has a wide range of applications including recommender systems, spam mail classification, identifying domain experts in various research areas, etc. In this chapter, we discuss the prior art in link prediction literature.

Keywords Social network · Undirected graph · Static · Dynamic link prediction · Local similarity measure · Heterogeneous network

Networks have become a part of our day-to-day lives with the advent of internet on smart phones, tablets, and computers. Any network can be represented as a graph $G = (V, E)$, where V corresponds to the set of nodes (users) of the network and E corresponds to set of links between the nodes (users). Online social networks like Twitter and Facebook allow a user to share messages, pictures, and videos with other users. Facebook may be viewed as an undirected network while Twitter can be viewed as a directed network. For example, the Facebook network can be represented as an undirected graph where nodes correspond to the users of Facebook and an edge exists between two nodes (users) if they are friends of each other on Facebook. Other networks can be represented similarly.

It is critical to understand how a network grows; how new nodes and links get added to the network with time. This problem is often referred to as the "network evolution problem." A detailed study of network evolution models can be found in [31]. However, in this book, we deal only with the *link prediction problem*, i.e., predicting new links that get added to the network over time; it does not consider addition or deletion of nodes to the network.

1.1 Link Prediction Problem

Given a network $G_t = (V, E_t)$ at a given time t, we need to predict the set of new links E which will most likely emerge in the network in the time interval $[t, t']$, where $t' > t$. The network $G_{t'}$ at time t' can be represented as $G_{t'} = (V, E_{t'})$ where

© The Author(s) 2016

V. Srinivas and P. Mitra, *Link Prediction in Social Networks*,
SpringerBriefs in Computer Science, DOI 10.1007/978-3-319-28922-9_1

$E_{t'} = E_t \cup E$. It is important to note that in the link prediction problem, V remains static with time. Contrastingly, in the network evolution problem, V varies over time as new nodes are added to the network.

1.2 Literature Review

The earliest work on link prediction [40, 52–54] was carried out on web networks for effective web page navigation. Markov models were used alongside the structure of the web to assist users to navigate from any given web page. Although the link prediction problem was previously studied, it was formally surveyed, for the first time, in [21]. It presents similarity measures[1] for estimating the similarity of nodes in a social network. The authors infer that information pertaining to future interactions can be extracted from network topology alone. We present a review of link prediction literature as follows:

- **Static Link Prediction**: Static link prediction problem considers *only a single snapshot of the network* for link prediction. Link prediction algorithms compute similarity of node pairs; similar nodes tend to connect to each other [41].
  - **Local Similarity Measures**: Local similarity measures compute the similarity between two nodes using *local neighborhood features like node degree, node neighbors, and common neighbors between the two nodes*. State-of-the-art local similarity measures for link prediction include **Preferential Attachment (PA)** [21], **Common Neighbors (CN)** [30], **Adamic Adar Index (AA)** [1], and **Resource Allocation Index (RA)** [51]. PA is the simplest and computationally the least inexpensive similarity measure which is based on the preferential attachment network evolution model [31]. According to PA, similarity between two nodes is directly proportional to the product of the degree of the two nodes. CN assigns a similarity value between two nodes that is directly proportional to the number of common nodes they share. AA and RA are weighted versions of CN. They compute similarity between two nodes as a weighted sum of the inverse of a function of the common nodes' degree. *The weights assigned to the common nodes are based on the power law degree distribution.* Hence, low-degree common nodes are given higher importance compared to higher degree common nodes in computing similarity.

    Link prediction similarity measures show significant improvement in denser social networks when using both the graph similarity measures and the weights of existing links in a social network in [28, 29]. Similarly, in [37], a link between two nodes is predicted based on the probability of information propagation between the two nodes. Further, link prediction and classification were

---

[1] Even though we use the term similarity measure, it is a similarity function and need not be a measure.

collectively studied in an algorithm in [4]. Lu and Zhou [26] experimentally show that weak ties play a significant role in the link prediction problem and emphasizing the contribution of the weak ties enhances link prediction performance. Leroy et al. [19] introduce the cold start link prediction problem of predicting the complete network structure when the network is totally missing while additional other information pertaining to the nodes is available. They propose a two-phase method based on a bootstrap probabilistic graph. The first phase generates an implicit network; the second phase applies probabilistic graph-based measures to produce the final prediction. Lee et al. [18] use a mathematical programming approach for predicting a future network utilizing the node degree distribution identified from historical observation of the past networks. They propose an integer programming problem to maximize the sum of the link scores respecting the node degree distribution of the networks. Cohen and Zohar [7] present an axiomatic framework based on property templates, which analyzes the relevance of vertices to the score and how removal of edges and vertices affects the score.

- **Global Similarity Measures**: Global similarity measures compute the similarity between two nodes using more comprehensive *global features such as the number of paths, information flow, etc. between two nodes*. Global similarity measures involve higher computation cost compared to local similarity measures. **Katz** measure [15] is a global similarity measure which computes the similarity based on number of paths between two nodes. It assigns larger weight to shorter paths and smaller weights to longer paths between two nodes. The major problem with Katz measure is the amount of computation involved in similarity computation. Another global measure is **Random Walk with Restart (RWR)** [32], which is based on the popular PageRank metric. RWR computes the similarity between two nodes as the probability of a random walker starting at the start node and emerging at the destination node at the steady state. PROPFLOW [24] is another global similarity measure based on the notion of information flow. It is very efficient when compared to the Katz and RWR similarity measures.

• **Dynamic Link Prediction**: Dynamic link prediction predicts new links *based on a stream of snapshots of the network over time*. As opposed to static link prediction, it does not consider a single snapshot of the network for predicting new links. Tylenda et al. [46] propose to incorporate the history information available from various snapshots over time for predicting new and recurrent links. Results show that incorporating time-stamps of past interactions significantly improves the link prediction performance. Rümmele et al. [36] approach solving link prediction problem by counting 3-node graphlets. Further work in this direction can be found in [20, 38].

• **Link Prediction in Heterogeneous Networks**: Data sparsity problem is a main challenge in link prediction tasks. Cao et al. [5] address this problem by jointly considering multiple heterogeneous link prediction tasks, which they refer to as the collective link prediction (CLP) problem. CLP problem is solved using a Bayesian framework which allows knowledge to be transferred adaptively while taking into account similarities between tasks. Related work on link prediction in heterogeneous networks can be found in [8, 10, 12, 17, 34].

- **Link Prediction in Signed Networks**: Symeonidis et al. [44] propose a time-efficient technique using a combination of local and global features for link prediction in signed networks. Further, significant improvement can be achieved considering the information from both positive and negative links. Chiang et al. [6] show a quantitative measure of a social network that can be used to derive a link prediction algorithm in signed networks. They present a supervised machine learning algorithm that uses features derived from longer cycles in the network; using these features enhances the performance against existing algorithms.

- **Unsupervised and Supervised-Based Learning Algorithms for Link Prediction**: Kashima and Abe [13] present an efficient incremental supervised learning algorithm based on a probabilistic model. It uses topological features of the network structure for predicting links between the nodes. Miller et al. [27] propose a nonparametric Bayesian technique to predict links in relational data. The approach simultaneously infers the number of features and also learns which entities have each feature. Link prediction performance in co-author network can be substantially enhanced by considering the dual graph obtained by projecting the original two-mode network over the set of publications alongside the co-authorship network [3]. Sarkar et al. [39] theoretically justify the success of some similarity measures using a similar class of graph generation models in which nodes are associated with locations in a latent metric space and connections are more likely between closer nodes. They also show bounds related to node's degree that plays an important role in link prediction, the relative importance of short paths versus long paths, and the effects of increasing nondeterminism in the link generation process on link prediction quality. Tasnádi and Berend [45] propose using implicit information from the restaurant review portal based on the ratings and languages used by the users. It uses supervised machine learning techniques to use the independent information given by the users to test the connectedness of the users. Further work in this direction can be found in [2, 9, 11, 16, 23, 25, 33, 42, 43].

- **Semi-Supervised Learning-Based Algorithms for Link Prediction**: A probability model for estimating the joint probability of occurrence of two nodes was proposed in [50]. The model when integrated with the existing network features improves the link prediction performance. A fast semi-supervised algorithm for link prediction was proposed in [14]. The algorithm predicts unknown parts of the network structure using auxiliary information such as node similarities and is applicable to multi-relational domains. Raymond and Kashima [35] propose fast and scalable algorithms compared to [14] for link propagation by introducing efficient procedures to solve large linear equations that appear in the method.

1.3 Background

As discussed previously, similar nodes tend to connect to each other. Hence, higher similarity between a pair of nodes implies higher chances of an edge emerging between the two nodes. In this section, we discuss some of the most popular link prediction similarity measures.

Notation

- V—vertices of the network,
- E_t—edges of the network at time t,
- $N(a)$—neighbors of node a,
- x_a—degree of node a.

1.3.1 Link Prediction Similarity Measures

In this section, we discuss the state-of-the-art link prediction similarity measures and illustrate the behavior of these similarity measures using the example network shown in Fig. 1.1. Consider computing similarity between nodes a_1 and b_1 in the figure. Further, let a and b be any two nodes.

1. **Preferential Attachment (PA)**

   Similarity between nodes a and b is calculated as the product of the degree of the nodes a and b. Specifically, we illustrate the similarity computation between nodes a and b in the example network. The higher the degree of both the nodes, the higher is the similarity between a and b:

   $$PA(a, b) = x_a \times x_b$$

Fig. 1.1 Example network for illustration

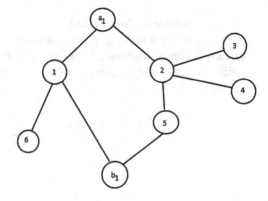

The similarity between nodes a_1 and b_1 using PA similarity measure can be computed as

$$PA(a_1, b_1) = x_{a_1} \times x_{b_1} = 2 \times 2 = 4$$

2. **Common Neighbors (CN)**
   Similarity between nodes a and b is calculated as the number of common neighbors between a and b. The higher the number of common nodes shared, the higher is the similarity between a and b:

$$CN(a, b) = \mid N(a) \cap N(b) \mid$$

The similarity between nodes a_1 and b_1 using CN similarity measure can be computed as

$$CN(a_1, b_1) = \mid N(a_1) \cap N(b_1) \mid = \mid \{1, 2\} \cap \{1, 5\} \mid = 1$$

3. **Adamic Adar Index (AA)**
   Similarity between nodes a and b is calculated as the sum of inverse of the logarithm of degree of each common neighbor z between a and b. The more the number of low degree common nodes shared, the higher is the similarity between a and b. This is a weighted version of the CN similarity measure:

$$AA(a, b) = \sum_{z \in N(a) \cap N(b)} \frac{1}{\log(x_z)}$$

The similarity between nodes a_1 and b_1 using AA similarity measure can be computed as

$$AA(a_1, b_1) = \sum_{z \in N(a_1) \cap N(b_1)} \frac{1}{\log(x_z)} = \frac{1}{\log(x_1)} = \frac{1}{\log(3)}$$

4. **Resource Allocation Index (RA)**
   Similarity between nodes a and b is calculated as the sum of inverse of the degree of each common neighbor z between a and b. This similarity measure is very similar to AA; it assigns lower weight to the higher degree common nodes when compared to AA:

$$RA(a, b) = \sum_{z \in N(a) \cap N(b)} \frac{1}{x_z}$$

The similarity between nodes a_1 and b_1 using RA similarity measure can be computed as

$$RA(a_1, b_1) = \sum_{z \in N(a_1) \cap N(b_1)} \frac{1}{x_z} = \frac{1}{x_1} = \frac{1}{3}$$

5. **Katz**

Similarity between nodes a and b is calculated based on the collection of all the paths damped by length to assign higher weights to shorter paths. β is a damping factor; using small values of β ignores longer length paths. $| \text{paths}_{a,b}^{(l)} |$ represents the number of paths of length exactly l between a and b. The higher the number of short-length paths, the higher is the similarity between a and b:

$$Katz(a, b) = \sum_{l=1}^{\infty} \beta^l \, | \text{paths}_{a,b}^{(l)} |$$

The similarity between nodes a_1 and b_1 using Katz similarity measure can be computed as

$$Katz(a_1, b_1) = \beta^2 \times | \text{paths}_{a,b}^{(2)} | + \beta^3 \times | \text{paths}_{a,b}^{(3)} |$$
$$= \beta^2 \times 1 + \beta^3 \times 1 = \beta^2 + \beta^3$$

where β is typically set to 0.005.

6. **PropFlow**

PropFlow is a similarity measure based on information flow. It represents the amount of information that flows from the source to the destination node across different paths. The higher the information gathered, the larger is the similarity value between source and destination node. The detailed PropFlow algorithm can be found in [24].

The similarity between nodes a_1 and b_1 using PropFlow similarity measure can be computed as the amount of flow at node b_1 with 1 unit flow starting at node a_1:

$$PropFlow(a_1, b_1) = flow(a_1, 1) + PropFlow(1, b_1) + flow(a_1, 2)$$
$$+ PropFlow(2, b_1)$$

where $flow(1, b_1)$ is the amount of flow from node 1 to node b_1:

$$flow(a_1, 1) = \frac{1}{x_{a_1}} = \frac{1}{2}$$

Performing similar operations recursively, we can find that

$$PropFlow(a_1, b_1) = \frac{1}{4} + \frac{1}{6}$$

where flow of $\frac{1}{4}$ comes from path $\{a_1, 1, b_1\}$ and flow of $\frac{1}{6}$ comes from path $\{a_1, 2, 5, b_1\}$.

1.3.2   Outline of Link Prediction Algorithm

In this section, we present an outline of how a link prediction similarity measure S can be used to predict the edges that will emerge in the network in the future as in [21].

Algorithm 1: Outline of Link Prediction Algorithm using S

---

input : Input Graph $G_t = (V, E_t)$
         lp is the number of likely edges we want to predict
         Similarity Measure S
output: Predicted Edges Based on Similarity
1 **for** $(a, b) \notin E_t$ **do**
2   compute the similarity using similarity measure S.
3 **end**
4 Sort the node pairs in descending order based on the computed score.
5 Output the the top lp links.

1.3.3   Power Law Degree Distribution

Power law degree distribution [31] can be defined as follows: The probability of finding a x degree node in the network, denoted by p_x, is directly proportional to $x^{-\alpha}$, where α is some positive constant. Hence, we can infer that the probability of finding a high degree node in the network is very small as the corresponding value of x is very high and the probability of finding a low degree node is relatively high as the corresponding value of x is small. We consider the power law degree distribution, as most of the large-scale networks can be approximated to follow a power law degree distribution. Further, we observe that AA and RA also obey the power law distribution by assigning smaller weight to the high-degree common neighbors.

1.3.4 Kullback–Leibler (KL) Divergence

We describe below the notion of KL divergence (KL) in the context of link prediction. Consider that graphs G_t and $G_{t'}$ ($t' > t$) have degree probability distributions q and p, respectively. Then, the KL divergence denoted by D_{KL} between p and q can be calculated as follows:

$$D_{KL}(p||q) = \sum_x p(x) \log \frac{p(x)}{q(x)}$$

We use KL divergence (D_{KL}) to measure the distance between two probability distributions. If D_{KL} is low, it means the distributions are very similar and vice versa. Analogously, we use KL divergence in the context of link prediction to measure the distance between degree distributions of the networks. Note that KL divergence function takes two arguments as input. However, in the rest of the book, we assume that one of the arguments is fixed; so, we explicitly indicate one argument as input. The fixed argument is the degree distribution of $G_{t'}$. Hence, KL(G) represents the KL divergence between the degree distributions of graph G and $G_{t'}$.

1.3.5 Clustering Coefficient

Clustering coefficient (CC) [31] is the average of the local CC of the nodes. The local CC of a node a is a fraction of the number of links between the neighbors of node a to the maximum possible number of links between them. Thus, when all the neighbors of the node a are not linked and are all linked, we attain the minimum and maximum values for $CC(a)$ as 0 and 1, respectively.

$$CC(a) = \frac{\text{\# links between neighbors of } a}{\text{maximum possible links between neighbors of } a}$$

A network has high value of CC when the CC value averaged across all the nodes is high. When a network has a large CC value, it implies that the network has large number of triangles (cliques of size 3). Due to the presence of large number of triangles, the nodes of the network satisfy a transitive relation which would enable a better performance of the similarity measures in link prediction. Hence, the similarity measures often do not provide good performance on sparse networks.

Table 1.1 Dataset statistics

Dataset	$\mid V \mid$	$\mid E \mid$	$\langle k \rangle$	CC	d
Amazon	334,863	925,872	5.529	0.3976	44
CondMat	23,133	93,497	8.083	0.63	14
HepTh	9877	25,998	5.26	0.4714	17

Table 1.2 Running Time (min)

	Amazon	CondMat	HepTh
CN	0.897	0.357	0.038
AA	0.9825	0.3695	0.0395
RA	1.053	0.3825	0.04
PA	0.133	0.0029	0.04
Katz	2000	478.5125	22.989
ProFlow	500.025	131.497	2.5935

1.4 Datasets Used in Experiments

We consider real-world networks that are provided by SNAP database (http://snap.
stanford.edu/data/). The details of the datasets are given in Table 1.1. The graph
datasets do not contain time-stamps representing the time at which links are formed
in the network. $\mid V \mid$, $\mid E \mid$, $\langle k \rangle$, CC, and d represent the number of nodes, number
of links, average degree, average clustering coefficient, and diameter of the network.
We can observe that Amazon network is much bigger (factor of 10) than HepTh
and CondMat networks in terms of size. However, Amazon and HepTh networks are
much sparser than the CondMat network.

All the experiments were conducted using the Boost graph library (http://www.
boost.org) in C++ on an Intel(R) machine with 2.2 GHz CPU and 8 GB RAM
running Windows 8.1. We adopt the C++ implementations of Katz and PropFlow
made available with the Lpmade package in [22]. The average running time (10
iterations) of various similarity measures for link prediction on various datasets is
shown in Table 1.2.

Observe that CN, AA, RA, and PA are the fastest similarity measures when compared to Katz and PropFlow. They take approximately 1 minute for running even
on the largest Amazon dataset. On the other hand, observe that Katz and PropFlow
take approximately 500–1000 times in running time when compared to CN, AA,
RA, and PA. On the smallest HepTh dataset, Katz and PropFlow take around 23 and
2.5 min running time, respectively. However, on the larger Amazon dataset, Katz and
Proflow take approximately 33 and 8 h running time, respectively. It can be observed
that these global measures are not scalable. Hence, from the results perspective, in
the rest of the book we do not consider Katz and PropFlow[2] similarities.

---

[2]Owing to high computational overhead.

References

1. Adamic, L.A., Adar, E.: Friends and neighbors on the web. Soc. Netw. **25**, 211–230 (2003)
2. Barbieri, N., Bonchi, F., Manco, G.: Who to follow and why: link prediction with explanations. In: The 20th ACM SIGKDD International Conference on Knowledge Discovery and Data Mining, KDD'14, pp. 1266–1275. New York, NY, USA, 24–27 August 2014
3. Benchettara, N., Kanawati, R., Rouveirol, C.: A supervised machine learning link prediction approach for academic collaboration recommendation. In: Proceedings of the 2010 ACM Conference on Recommender Systems, RecSys 2010, pp. 253–256. Barcelona, Spain, 26–30 September 2010
4. Bilgic, M., Namata, G., Getoor, L.: Combining collective classification and link prediction. In: Workshops Proceedings of the 7th IEEE International Conference on Data Mining (ICDM 2007), pp. 381–386. Omaha, Nebraska, 28–31 October 2007
5. Cao, B., Liu, N.N., Yang, Q.: Transfer learning for collective link prediction in multiple heterogenous domains. In: Proceedings of the 27th International Conference on Machine Learning (ICML-10), pp. 159–166. Haifa, Israel, 21–24 June 2010
6. Chiang, K., Natarajan, N., Tewari, A., Dhillon, I.S.: Exploiting longer cycles for link prediction in signed networks. In: Proceedings of the 20th ACM Conference on Information and Knowledge Management, CIKM 2011, pp. 1157–1162. Glasgow, United Kingdom, 24–28 October 2011
7. Cohen, S., Zohar, A.: An axiomatic approach to link prediction. In: Proceedings of the Twenty-Ninth AAAI Conference on Artificial Intelligence, pp. 58–64. Austin, Texas, USA, 25–30 January 2015
8. Davis, D.A., Lichtenwalter, R., Chawla, N.V.: Multi-relational link prediction in heterogeneous information networks. In: International Conference on Advances in Social Networks Analysis and Mining, ASONAM 2011, pp. 281–288. Kaohsiung, Taiwan, 25–27 July 2011
9. De, A., Ganguly, N., Chakrabarti, S.: Discriminative link prediction using local links, node features and community structure. In: 2013 IEEE 13th International Conference on Data Mining, pp. 1009–1018. Dallas, TX, USA, 7–10 December 2013
10. Dong, Y., Tang, J., Wu, S., Tian, J., Chawla, N.V., Rao, J., Cao, H.: Link prediction and recommendation across heterogeneous social networks. In: 12th IEEE International Conference on Data Mining, ICDM 2012, pp. 181–190. Brussels, Belgium,10–13 December 2012
11. Gao, S., Denoyer, L., Gallinari, P.: Link prediction via latent factor blockmodel. In: Proceedings of the 21st World Wide Web Conference, WWW 2012 (Companion Volume), pp. 507–508. Lyon, France, 16–20 April 2012
12. Ge, L., Zhang, A.: Pseudo cold start link prediction with multiple sources in social networks. In: Proceedings of the Twelfth SIAM International Conference on Data Mining, pp. 768–779. Anaheim, California, USA, 26–28 April 2012
13. Kashima, H., Abe, N.: A parameterized probabilistic model of network evolution for supervised link prediction. In: Proceedings of the 6th IEEE International Conference on Data Mining (ICDM 2006), pp. 340–349. Hong Kong, China, 18–22 December 2006
14. Kashima, H., Kato, T., Yamanishi, Y., Sugiyama, M., Tsuda, K.: Link propagation: A fast semi-supervised learning algorithm for link prediction. In: Proceedings of the SIAM International Conference on Data Mining, SDM 2009, pp. 1100–1111. Sparks, Nevada, USA, 30 April–2 May 2009
15. Katz, L.: A new status index derived from sociometric analysis. Psychometrika **18**(1), 39–43 (1953)
16. Kim, J., Choy, M., Kim, D., Kang, U.: Link prediction based on generalized cluster information. In: 23rd International World Wide Web Conference, WWW'14 (Companion Volume), pp. 317–318. Seoul, Republic of Korea, 7–11 April 2014
17. Kuo, T., Yan, R., Huang, Y., Kung, P., Lin, S.: Unsupervised link prediction using aggregative statistics on heterogeneous social networks. In: The 19th ACM SIGKDD International Conference on Knowledge Discovery and Data Mining, KDD 2013, pp. 775–783. Chicago, IL, USA, 11–14 August 2013

18. Lee, C., Pham, M., Kim, N., Jeong, M.K., Lin, D.K.J., Chaovalitwongse, W.A.: A novel link prediction approach for scale-free networks. In: 23rd International World Wide Web Conference, WWW'14 (Companion Volume), pp. 1333–1338. Seoul, Republic of Korea, 7–11 April 2014

19. Leroy, V., Cambazoglu, B.B., Bonchi, F.: Cold start link prediction. In: Proceedings of the 16th ACM SIGKDD International Conference on Knowledge Discovery and Data Mining, pp. 393–402. ACM (2010)

20. Li, X., Du, N., Li, H., Li, K., Gao, J., Zhang, A.: A deep learning approach to link prediction in dynamic networks. In: Proceedings of the 2014 SIAM International Conference on Data Mining, pp. 289–297. Philadelphia, Pennsylvania, USA, 24–26 April 2014

21. Liben-Nowell, D., Kleinberg, J.M.: The link prediction problem for social networks. In: Proceedings of the 2003 ACM CIKM International Conference on Information and Knowledge Management, pp. 556–559. New Orleans, Louisiana, USA, 2–8 November 2003

22. Lichtenwalter, R., Chawla, N.V.: Lpmade: Link prediction made easy. J. Mach. Learn. Res. **12**, 2489–2492 (2011)

23. Lichtenwalter, R., Chawla, N.V.: Vertex collocation profiles: subgraph counting for link analysis and prediction. In: Proceedings of the 21st World Wide Web Conference 2012, WWW 2012, pp. 1019–1028. Lyon, France, 16–20 April 2012

24. Lichtenwalter, R.N., Lussier, J.T., Chawla, N.V.: New perspectives and methods in link prediction. In: Proceedings of the 16th ACM SIGKDD International Conference on Knowledge Discovery and Data Mining—KDD'10, p. 243 (2010)

25. Liu, F., Liu, B., Wang, X., Liu, M., Wang, B.: Features for link prediction in social networks: A comprehensive study. In: Proceedings of the IEEE International Conference on Systems, Man, and Cybernetics, SMC 2012, pp. 1706–1711. Seoul, Korea (South), 14–17 October 2012

26. Lu, L., Zhou, T.: Role of weak ties in link prediction of complex networks. In: Proceeding of the ACM First International Workshop on Complex Networks Meet Information & Knowledge Management, CIKM-CNIKM 2009, pp. 55–58. Hong Kong, China, 6 November 2009

27. Miller, K.T., Griffiths, T.L., Jordan, M.I.: Nonparametric latent feature models for link prediction. In: Advances in Neural Information Processing Systems 22: 23rd Annual Conference on Neural Information Processing Systems 2009, pp. 1276–1284. Vancouver, British Columbia, Canada, 7–10 December 2009

28. Murata, T., Moriyasu, S.: Link prediction of social networks based on weighted proximity measures. In: 2007 IEEE/WIC/ACM International Conference on Web Intelligence, WI 2007, Main Conference Proceedings, pp. 85–88. Silicon Valley, CA, USA, 2–5 November 2007

29. Murata, T., Moriyasu, S.: Link prediction based on structural properties of online social networks. New Gener. Comput. **26**(3), 245–257 (2008)

30. Newman, M.E.J.: Clustering and preferential attachment in growing networks. Phys. Rev. E, Stat. Nonlinear Soft Matter Phys. **64**, 025102 (2001)

31. Newman, M.E.J.: Networks: An Introduction. Oxford University Press Inc, New York (2010)

32. Pan, J.Y., Yang, H.J., Faloutsos, C., Duygulu, P.: Automatic multimedia cross-modal correlation discovery. In: Proceedings of the Tenth ACM SIGKDD International Conference on Knowledge Discovery and Data Mining, pp. 653–658. ACM (2004)

33. Pujari, M., Kanawati, R.: Supervised rank aggregation approach for link prediction in complex networks. In: Proceedings of the 21st World Wide Web Conference, WWW 2012 (Companion Volume), pp. 1189–1196. Lyon, France, 16–20 April 2012

34. Qi, G., Aggarwal, C.C., Huang, T.S.: Link prediction across networks by biased cross-network sampling. In: 29th IEEE International Conference on Data Engineering, ICDE 2013, pp. 793–804. Brisbane, Australia, 8–12 April 2013

35. Raymond, R., Kashima, H.: Fast and scalable algorithms for semi-supervised link prediction on static and dynamic graphs. In: Machine Learning and Knowledge Discovery in Databases, European Conference, ECML PKDD 2010, Proceedings, Part III, pp. 131–147. Barcelona, Spain, 20–24 September 2010

36. Rümmele, N., Ichise, R., Werthner, H.: Exploring supervised methods for temporal link prediction in heterogeneous social networks. In: Proceedings of the 24th International Conference on World Wide Web Companion, WWW 2015 (Companion Volume), pp. 1363–1368. Florence, Italy, 18–22 May 2015

37. Saito, K., Nakano, R., Kimura, M.: Prediction of link attachments by estimating probabilities of information propagation. In: Knowledge-Based Intelligent Information and Engineering Systems, 11th International Conference, KES 2007, XVII Italian Workshop on Neural Networks, Proceedings, Part III, pp. 235–242. Vietri sul Mare, Italy, 12–14 September 2007

38. Sarkar, P., Chakrabarti, D., Jordan, M.I.: Nonparametric link prediction in dynamic networks. In: Proceedings of the 29th International Conference on Machine Learning, ICML 2012, Edinburgh, Scotland, UK, 26 June–1 July 2012

39. Sarkar, P., Chakrabarti, D., Moore, A.W.: Theoretical justification of popular link prediction heuristics. In: IJCAI 2011, Proceedings of the 22nd International Joint Conference on Artificial Intelligence, pp. 2722–2727. Barcelona, Catalonia, Spain, 16–22 July 2011

40. Sarukkai, R.: Link prediction and path analysis using markov chains. Comput. Netw. **33**(1–6), 377–386 (2000)

41. Scholz, M.: Node similarity as a basic principle behind connectivity in complex networks (2010). ArXiv preprint arXiv:1010.0803

42. Shin, D., Si, S., Dhillon, I.S.: Multi-scale link prediction. In: 21st ACM International Conference on Information and Knowledge Management, CIKM'12, pp. 215–224. Maui, HI, USA, 29 October–02 November 2012

43. Soundarajan, S., Hopcroft, J.E.: Using community information to improve the precision of link prediction methods. In: Proceedings of the 21st World Wide Web Conference, WWW 2012 (Companion Volume), pp. 607–608. Lyon, France, 16–20 April 2012

44. Symeonidis, P., Tiakas, E., Manolopoulos, Y.: Transitive node similarity for link prediction in social networks with positive and negative links. In: Proceedings of the 2010 ACM Conference on Recommender Systems, RecSys 2010, pp. 183–190. Barcelona, Spain, 26–30 September 2010

45. Tasnádi, E., Berend, G.: Supervised prediction of social network links using implicit sources of information. In: Proceedings of the 24th International Conference on World Wide Web Companion, WWW 2015 (Companion Volume), pp. 1117–1122. Florence, Italy, 18–22 May 2015

46. Tylenda, T., Angelova, R., Bedathur, S.J.: Towards time-aware link prediction in evolving social networks. In: Proceedings of the 3rd Workshop on Social Network Mining and Analysis, SNAKDD 2009, p. 9. Paris, France, 28 June 2009

47. Virinchi, S., Mitra, P.: Link prediction using power law clique distribution and common edges distribution. In: Proceedings of Pattern Recognition and Machine Intelligence—5th International Conference, PReMI 2013, pp. 739–744. Kolkata, India, 10–14 December 2013

48. Virinchi, S., Mitra, P.: Similarity measures for link prediction using power law degree distribution. In: Proceedings of Neural Information Processing—20th International Conference, ICONIP 2013, Part II, pp. 257–264. Daegu, Korea, 3–7 November 2013

49. Virinchi, S., Mitra, P.: Two-phase approach to link prediction. In: Proceedings of Neural Information Processing—21st International Conference, ICONIP 2014, Part II, pp. 413–420. Kuching, Malaysia, 3–6 November 2014

50. Wang, C., Satuluri, V., Parthasarathy, S.: Local probabilistic models for link prediction. In: Proceedings of the 7th IEEE International Conference on Data Mining (ICDM 2007), pp. 322–331. Omaha, Nebraska, USA, 28–31 October 2007

51. Zhou, T., Lü, L., Zhang, Y.C.: Predicting missing links via local information. Eur. Phys. J. B **71**, 623–630 (2009)

52. Zhu, J.: Using markov chains for structural link prediction in adaptive web sites. In: Proceedings of User Modeling 2001, 8th International Conference, UM 2001, pp. 298–302. Sonthofen, Germany, 13–17 July 2001

53. Zhu, J., Hong, J., Hughes, J.G.: Using markov chains for link prediction in adaptive web sites. In: Proceedings of Soft-Ware 2002: Computing in an Imperfect World, First International Conference, Soft-Ware 2002, pp. 60–73, Belfast, Northern Ireland, 8–10 April 2002

54. Zhu, J., Hong, J., Hughes, J.G.: Using markov models for web site link prediction. In: Proceedings of the 13th ACM Conference on Hypertext and Hypermedia, HYPERTEXT 2002, pp. 169–170. University of Maryland, College Park, MD, USA, 11–15 June 2002

Chapter 2
Link Prediction Using Thresholding Nodes Based on Their Degree

Abstract In this chapter, we propose MIDT, a degree threshold-based similarity measure, for link prediction which exploits the power law degree distribution which social networks typically follow. We show that, in power law networks, the number of high-degree common neighbors is insignificant compared to the low-degree common neighbors. We use this property to assign a zero weight to the high-degree common neighbors and a higher weight to the low-degree neighbors in computing similarity between nodes. Experiments on standard benchmark datasets show the superiority of MIDT similarity measure. Specifically, MIDT shows an improvement of upto 4 % in terms of AUC when compared to the state-of-the-art link prediction similarity measures.

Keywords Power law degree distribution · Markov inequality · Degree thresholding · Clique-based approach · Area under the curve (AUC)

2.1 Introduction

Most networks can be approximated to follow the power law degree distribution. Accordingly, the probability of encountering a high-degree node is very small. Similarly, the frequency of distinct terms in large-scale collection of documents follows the Zipfian distribution which is a simple form of the power law degree distribution. In Information Retrieval, in the well-known technique called *stopping*, which is used for indexing document collection, the terms with high-frequency are often removed or not considered for indexing purpose as the high-frequency terms often lack good discriminating capability.

Analogously, it is possible that *high-degree nodes lack good discriminating capability; each high-degree node is connected to many other nodes*. Hence, we will need a threshold value on the degree in order to classify a node into low-degree and high-degree clusters. We use an appropriate threshold to split the set of nodes based on their degree into low-degree and high-degree node sets. Once, we classify the nodes into

---

Material in this chapter appeared in [6].

© The Author(s) 2016

V. Srinivas and P. Mitra, *Link Prediction in Social Networks*,

SpringerBriefs in Computer Science, DOI 10.1007/978-3-319-28922-9_2

low-degree and high-degree node sets, then, we give different weights to common
nodes in different clusters while computing the similarity between a pair of unconnected nodes. We emphasize the role of low-degree common nodes in terms of their
contribution to the similarity and ignore the contribution of high-degree common
nodes. We formally show that the number of high-degree common nodes between
any pair of nodes is very small. Hence, we justify the proposed scheme formally which
de-emphasizes the role of high-degree common nodes in computing similarity
between two nodes. The modified similarity measure shows an improved performance on the benchmark datasets.

2.2 Markov Inequality for Determining Threshold

In general, given G_t, it is nontrivial to obtain a threshold for classifying nodes into low-
degree and high-degree clusters. We use Markov inequality to derive a suitable value
for this threshold. Markov inequality can be stated as follows: if X is a nonnegative
random variable and there exists some positive constant $r > 0$,

$$P(X \geq r) \leq \frac{E[X]}{r}$$

We use the above inequality to obtain a threshold value (T) for dividing the set of
nodes into low-degree and high-degree clusters. In this case, we take degree as the
non-negative random variable (degree of a node is always nonnegative) and T will
be positive. We can rewrite for the required inequality as:

$$P(degree \geq T) \leq \frac{E[degree]}{T} \Rightarrow T \leq \frac{E[degree]}{P(degree \geq T)}. \tag{2.1}$$

Thus, from Eq. 2.1 we can calculate the required threshold based on the number
of high-degree nodes that we can ignore. We use Eq. 2.1 to bound the threshold
value. For conducting experiments, we set $P(degree \geq T)$ to 0.1. Using the calculated
threshold T, we divide the node set V into low-degree and high-degree node sets.
We justify our de-emphasis of the high-degree common neighbors using the theorem
below.

Notation

- x_y—degree of node y
- p_k—probability of existence of a k degree node in the graph
- n—number of nodes in the graph
- n_L—number of low-degree nodes, n_H—Number of high-degree nodes
- L—low-degree cluster $\{y | x_y < T\}$, H—High-degree cluster $\{y | x_y \geq T\}$
- L_{avg}—average degree of a node in L, H_{avg}—Average degree of a node in H

- K_L—expected number of low-degree common neighbors between any pair of nodes
- K_H—expected number of high-degree common neighbors between any pair of nodes
- T—degree Threshold, *max*—Maximum degree of a node in the network

Theorem 2.1 *For any pair of nodes a and b, the expected number of high-degree common neighbors (K_H) is very small when compared to expected number of low-degree common neighbors (K_L), i.e., $K_H \ll K_L$.*

Proof Consider the possibility that both $a, b \in L$. The probability that a common neighbor $z \in L$ is given by

$$P(z \in L) = \frac{n_L}{n} \times p_{L_{avg}} \times \frac{n_L}{n} \times p_{L_{avg}} \times \frac{n_L}{n} \times p_{L_{avg}} \times \frac{n_L}{n} \times p_{L_{avg}} \qquad (2.2)$$

where $\frac{n_L}{n}$ accounts for the selection of a node from L and $p_{L_{avg}}$ accounts for the average probability of existence of a low-degree node. Note that the first four terms correspond to the probability of existence of a link between a and z and the last four terms correspond to the probability of existence of a link between b and z.

The above equation can be simplified to the following form:

$$P(z \in L) = \frac{n_L}{n} \times p_{L_{avg}} \times \frac{n_L}{n} \times p_{L_{avg}} \times \left(\frac{n_L}{n} \times p_{L_{avg}} \right)^2 \qquad (2.3)$$

In a similar way the probability that a common neighbor $z \in H$ is given by

$$P(z \in H) = \frac{n_L}{n} \times p_{L_{avg}} \times \frac{n_L}{n} \times p_{L_{avg}} \times \left(\frac{n_H}{n} \times p_{H_{avg}} \right)^2 \qquad (2.4)$$

So,

$$K_L = n \times P(z \in L) = n \times \frac{n_L}{n} \times p_{L_{avg}} \times \frac{n_L}{n} \times p_{L_{avg}} \times \left(\frac{n_L}{n} \times p_{L_{avg}} \right)^2 \qquad (2.5)$$

and

$$K_H = n \times P(z \in H) = n \times \frac{n_L}{n} \times p_{L_{avg}} \times \frac{n_L}{n} \times p_{L_{avg}} \times \left(\frac{n_H}{n} \times p_{H_{avg}} \right)^2 \qquad (2.6)$$

Thus, the ratio of K_H to K_L, from Eqs. 2.5 and 2.6 is given by

$$\frac{K_H}{K_L} = \left(\frac{n_H}{n_L} \right)^2 \times \left(\frac{p_{H_{avg}}}{p_{L_{avg}}} \right)^2 \qquad (2.7)$$

a and b can be assigned to L and H in 3 other ways as follows:

1. $a \in L$ and $b \in H$
2. $a \in H$ and $b \in L$
3. $a \in H$ and $b \in H$

Note that in all these 3 cases also, the value of $\frac{K_H}{K_L}$ is the same as the one given in Eq. 2.7. Using power law degree distribution property of the network,

$$p_{L_{avg}} = C \times (L_{avg})^{-\alpha} \tag{2.8}$$

$$\text{and, } p_{H_{avg}} = C \times (H_{avg})^{-\alpha}$$

From Eqs. 2.7 and 2.8, we have

$$\frac{K_H}{K_L} = \left(\frac{n_H}{n_L}\right)^2 \times \left(\frac{L_{avg}}{H_{avg}}\right)^{2\alpha} \tag{2.9}$$

Consider $\frac{n_H}{n_L}$ which can be simplified as,

$$\frac{n_H}{n_L} = \frac{n \times \sum_{i=T}^{max} p_i}{n \times \sum_{i=1}^{T-1} p_i} \tag{2.10}$$

Note that

$$2^{-\alpha} \leq \sum_{i=1}^{T} i^{-\alpha} \leq (T-1) \times 2^{-\alpha} \tag{2.11}$$

Using power law we can substitute $i^{-\alpha}$ for p_i and cancelling out n, we get

$$\frac{n_H}{n_L} = \frac{\sum_{i=T}^{max} i^{-\alpha}}{\sum_{i=1}^{T-1} i^{-\alpha}} \leq \frac{(max - T) \times T^{-\alpha}}{2^{-\alpha}} = (max - T) \times \left(\frac{2}{T}\right)^{\alpha} \tag{2.12}$$

Also by noting that $L_{avg} < T$ and $H_{avg} > T$ we can bound $\frac{L_{avg}}{H_{avg}}$ as follows,

$$\frac{L_{avg}}{H_{avg}} = \frac{T - \delta}{T + \delta} \quad \text{for some } 1 < \delta < T \tag{2.13}$$

$$\frac{L_{avg}}{H_{avg}} = \left(1 - \frac{2\delta}{T + \delta}\right) < e^{-\left(\frac{2\delta}{T+\delta}\right)} \tag{2.14}$$

So from 2.9, 2.12 and 2.14, we get

$$\frac{K_H}{K_L} \leq (max - T)^2 \times \left(\frac{2}{T}\right)^{2\alpha} \times e^{-\left(\frac{2\delta}{T+\delta}\right)2\alpha} \tag{2.15}$$

which is a very small quantity and it tends to 0 as T tends to a large value which happens when the graph is large. It is intuitively clear that $L_{avg} < H_{avg}$. Further, because of the power law and selection of an appropriate threshold value we can make $\frac{n_H}{n_L}$ as small as possible. For example, by selecting the value of T to be less than or equal to $\frac{max}{2}$ we get $\frac{n_H}{n_L}$ to range between 0.003 and 0.04 for several benchmark datasets and for the same threshold the value of $\frac{L_{avg}}{H_{avg}}$ ranges from 0.07 to 0.14. So, the value of $\frac{K_H}{K_L}$ ranges from $0.000009 * (0.0049)^\alpha$ to $0.019 * (0.0016)^\alpha$. Typically, the value of α lies between 2 and 3. So, $\frac{K_H}{K_L}$ can be very small. Thus,

$$K_H \ll K_L \qquad \qquad \square$$

2.3 Markov Inequality-Based Degree Thresholding Similarity Measure (MIDT)

We present the proposed similarity measure (MIDT) which is based on degree thresholding of nodes. The proposed similarity measure assigns a higher weight to the low-degree common neighbors and a zero weight to the high-degree common neighbors. Specifically, this idea is similar to *stopping*, which is used for indexing document collection; the terms with high-frequency are often removed or not considered for indexing purpose as the high-frequency terms often lack good discriminating capability. Let us represent the weight assigned to a common node z by $w(z)$. As discussed before, we can observe that PA, CN, AA, and RA maintain weight $w(z)$ of 0, 1, $\frac{1}{\log(x_z)}$ and $\frac{1}{x_z}$, respectively. We show the variation of weight assigned to a common node z using various similarity measures in Fig. 2.1.

Observe from Fig. 2.1 that the weight assigned to a common node z by AA and RA fall very steeply with degree and become negligible for high-degree common

Fig. 2.1 Weight of a common node z using various similarity measures

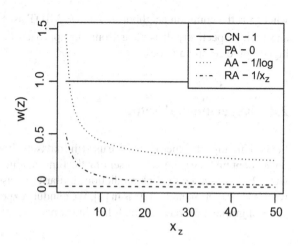

nodes. Now we present **MIDT** similarity measure. Let $N(a)$ and $N(b)$ represent the set of neighbors of nodes a and b, respectively. Further, let $R = N(a) \cap N(b)$.

$$MIDT(a, b) = \sum_{z \in R \wedge x_z < T} \frac{1}{\sqrt{x_z}} \qquad (2.16)$$

Next, we present the outline of the link prediction algorithm using MIDT similarity measure.

Algorithm 2: Outline of Link Prediction Algorithm using MIDT

---

   input : Input Graph $G_t = (V, E_t)$
           lp is the number of likely edges we want to predict
   output: Predicted Edges Based on Similarity
1 **for** $(a, b) \notin E_t$ **do**
2

$$MIDT(a, b) = \sum_{z \in R \wedge x_z < T} \frac{1}{\sqrt{x_z}}$$

3 **end**
4 Sort the node pairs in descending order based on the computed MIDT score.
5 Output the the top lp links.

In general, we can write the similarity score function as a combination of two monotonically nonincreasing functions *low* and *high*, where *low* is applied on common neighbors having degree less than threshold and *high* is applied on common neighbors having degree greater than the threshold.

$$score(a, b) = \sum_{z \in R} (low(z) + high(z)) \qquad (2.17)$$

where z is the common neighbor of a and b. MIDT uses $\frac{1}{\sqrt{x_z}}$ and 0 for functions low and high respectively. It is clear from our approach that we assign zero weight to high-degree common nodes.

2.4   Experimental Setup

For conducting our experiments, we used the datasets shown in Table 1.1. The datasets do not contain time stamps representing the time at which links are formed in the network. For evaluating the link prediction similarity measures on such graph datasets, we use the experimental setup as in [3]. We conduct experiments on sampled training and test graphs generated in the following ways:

1. Perform edge sampling to divide the dataset into two parts each having 50 % links. We use one part as the training graph (G_t) and the other part as the test graph $(G_{t'})$. We predict the edges of $G_{t'}$. Let us call this **50–50 edge sampling**. 50–50 edge sampling indicates that training graph G_t has 50 % links and test graph $G_{t'}$ has the remaining 50 % links. This simulates link prediction on dense networks.
2. Perform edge sampling to divide the dataset into two parts each having 80 and 20 % links. We use the larger part as the training graph (G_t) and the smaller part as the test graph $(G_{t'})$. We predict the edges of $G_{t'}$. Let us call this **80–20 edge sampling**. 80–20 edge sampling indicates that training graph G_t has 80 % links and test graph $G_{t'}$ has the remaining 20 % links. This simulates link prediction on dense networks.

We use randomly sampled G_t as the graph at the current time instance and predict the links of the test graph $(G_{t'})$ to validate the predictions. We repeat this process 10 times to reduce any statistical bias introduced due to sampling.

For evaluating the performance of the link prediction similarity measures, we present the results in terms of the Area Under Curve (AUC) metric. While accuracy, precision, recall, and top-k equivalents are commonly used in link prediction literature, they are unstable due to class imbalance that arises in the link prediction problem [2]. We use the robust AUC metric which is stable and measures the area under the ROC curve for evaluating link predictor performance.

AUC, in the context of link prediction, can be computed as explained. Consider n random experiments of picking a correctly classified edge and a misclassified edge, if n_1 is the number of times the correctly classified edge has a higher score than the misclassified edge and n_2 is the number of times both have the same score. Then, AUC score can be computed as:

$$\text{AUC} = \frac{n_1 + 0.5 \times n_2}{n}$$

For our experiment we choose n to be 10000. AUC score indicates the ranking of the edges predicted based on the link prediction similarity measure. Higher AUC value implies that the similarity measure is a better ranking algorithm and yields better recommendations. We discuss the results in the next section.

2.5 Results

On performing the experiments using the MIDT similarity measures we report the AUC results in Tables 2.1 and 2.2 on various datasets on predicting 50 and 20 % missing links respectively.

From Tables 2.1 and 2.2, we observe that MIDT performs better than PA, CN, AA, and RA. We show the best results on each dataset when performing 50–50 and 80–20 edge sampling in boldface. Note that G_t is more dense in the case of 80–20

Table 2.1 AUC results for 50–50 edge sampling

Dataset	PA	CN	AA	RA	MIDT
Amazon	11.49	29.91	69.81	68.47	**71.37**
CondMat	35.59	65.15	73.53	68.35	**73.82**
HepTh	49.67	59.46	68.12	66.21	**68.82**

Table 2.2 AUC results for 80–20 edge sampling

Dataset	PA	CN	AA	RA	MIDT
Amazon	8.8	29.73	45.97	43.64	**48.52**
CondMat	55.01	64.42	79.34	81.22	**82.78**
HepTh	43.34	57.06	74.29	72.64	**78.07**

edge sampling when compared to 50–50 edge sampling. Hence, we can note that the link prediction similarity measures show a better performance of AUC in the case of 80–20 edge sampling when compared to 50–50 edge sampling.

Further, note that in general the AUC results of various link prediction similarity measures show better performance on the CondMat dataset when compared to the Amazon and HepTh dataset. The performance of link prediction similarity measure deteriorates with sparsity of training graph (G_t) [1]; CC is a good indicator of the density of a graph.

We can observe that MIDT performs the best when compared to the existing similarity measures in both the cases of 50–50 and 80–20 edge sampling. Further, MIDT performs better in the case of 80–20 edge sampling when compared to 50–50 edge sampling. In other words, MIDT shows better performance when the training graph is dense. Note that the AUC performance has increased by up to 4 %.

2.6 Clique-Based Approach to Link Prediction

Social networks follow the power law degree distribution [4]. We can exploit the power law degree distribution by ignoring the contribution of edges between a pair of high-degree nodes.[1] We consider the role of common edges between nodes where one of them is a low-degree node.

A clique in a graph G is any completely connected subgraph of G. In our method, we use cliques of size 3 to find out the common edges between two unconnected nodes. Figure 2.2 shows a common edge (u, v) shared by nodes a and b. By common edge (u, v) between two nodes a and b, we refer to the edge that is common to both the cliques auv and buv of size 3. Further, notice that there can be at most one edge common to two cliques of size 3. Note that we are considering cliques of size 3 to find the common edges and we ignore large size cliques as they are infrequent according to the power law distribution of cliques.

---

[1] Material in this chapter appeared in [5].

Fig. 2.2 Common edge
shared by nodes a and b

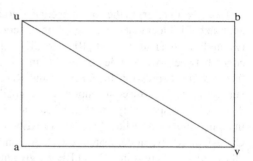

In the proposed algorithm, we make use of cliques for link prediction. The proposed similarity measures compute the similarity between two nodes by computing the similarity using the common node information and common edge information which we propose by making use of cliques of size 3. Specifically, for common edges shared we give higher weight to the edge if the endpoints of the common edge are low-degree nodes. Note that to define the threshold T, we use the same approach as explained before using Markov Inequality. If the endpoints of the common edge are high-degree nodes then the similarity contribution from the common edge is 0. Now, we present the proposed similarity measures which take into account the common edge information to the existing similarity measures. Let us call them **CN Using Cliques (CNC)**, **AA Using Cliques (AAC)**, and **RA Using Cliques (RAC)** which represent the extension for the existing similarity measures CN, AA, and RA, respectively.

$$CNC(a, b) = CN(a, b) + CE(a, b), \text{ where}$$

$$CE(a, b) = \begin{cases} 0 & \text{if } x_u > T \text{ and } x_v > T \\ \sum_{(u,v)} 2 & \text{else} \end{cases}$$

$$AAC(a, b) = AA(a, b) + AE(a, b), \text{ where}$$

$$AE(a, b) = \begin{cases} 0 & \text{if } x_u > T \text{ and } x_v > T \\ \sum_{(u,v)} 2/(\log(x_u) + \log(x_v)) & \text{else} \end{cases}$$

$$RAC(x, y) = RA(x, y) + RE(x, y), \text{ where}$$

$$RE(x, y) = \begin{cases} 0 & \text{if } degree(x) > T \text{ and } x_v > T \\ \sum_{(u,v)} 2/(x_u + x_v) & \text{else} \end{cases}$$

Here, (u, v) refers to the common edge shared by a cliques auv and buv. From our approach it is clear that we are ignoring the contribution of common edges between two high-degree nodes (in AE, RE, and CE) and giving more importance to common edges between two low-degree nodes (in AE and RE). Here, we choose two as a factor in the expression for CE, AE and RE as each common edge has two end vertices u and v whose contribution are weighed in a manner similar to to that of CN, AA and RA respectively. In other words, as there can be one-degree nodes present in G_t the maximum value of CE, AE or RE will be 1. This avoids the domination of CE, AE or RE when compared to CN, AA and RA. In CE, we give equal score to all the edges whereas in AE and RE we give more importance to the edge if one of the end vertices u and v is of low degree which varies as $2/(x_u + x_v)$. The CE, AE, and RE score is zero if the end vertices u and v both have degrees greater than the threshold T.

The results can be found in [5]. From the results, we can conclude that clique-based similarity measures perform better than the original similarity measures in terms of AUC. Also common edges between high-degree nodes are not so useful in predicting new links. Thus, we completely ignore the contributions of common edges between high-degree nodes.

2.7 Summary

In this chapter, we proposed the MIDT similarity measure which is based on thresholding nodes based on their degree. Experiments show that MIDT performs better than PA, CN, AA, and RA by upto 4 % in terms of AUC. Further, it can decrease time when the contribution of high-degree neighbors is ignored. The MIDT framework shows that low-degree common neighbors matter the most for link prediction and the high-degree neighbors can be ignored in predicting new links. We can completely ignore or minimize the contributions of high-degree nodes by making use of a suitable nonlinear similarity measure to weigh their contributions accordingly. In the future, we would like to consider other tight bounds to learn the threshold value (T); Markov Inequality gives a loose upper bound. Further, we can concentrate on constructing suitable nonlinear similarity measures. In the next chapter, we present a generic form of the existing similarity measures like CN, AA, and RA which uses the degree distribution of the network explicitly.

We can conclude that edge-based similarity measures perform better than the existing similarity measures in terms of AUC. Also common edges between high-degree nodes are not so useful in predicting new links. Thus, we completely ignore the contributions of common edges between high-degree nodes. We would like to design better similarity schemes that exploit the power law distribution of clique sizes; we would like to consider the contributions of bigger size (more than size 3) cliques.

References

1. Feng, X., Zhao, J., Xu, K.: Link prediction in complex networks: a clustering perspective. Eur. Phys. J. B **85**(1), 3 (2012)
2. Lichtenwalter, R., Chawla, N.V.: Link prediction: fair and effective evaluation. In: Proceedings of the 2012 IEEE/ACM International Conference on Advances in Social Networks Analysis and Mining, ASONAM 2012, pp. 376–383 (2012). doi:10.1109/ASONAM.2012.68
3. Lü, L., Zhou, T.: Link prediction in complex networks: a survey. Phys. A **390**(6), 1150–1170 (2011)
4. Newman, M.E.J.: Networks: An Introduction. Oxford University Press Inc, New York (2010)
5. Virinchi, S., Mitra, P.: Link prediction using power law clique distribution and common edges distribution. In: Proceedings of the Pattern Recognition and Machine Intelligence—5th International Conference, PReMI 2013, pp. 739–744. Kolkata, India, 10–14 December 2013
6. Virinchi, S., Mitra, P.: Similarity measures for link prediction using power law degree distribution. In: Proceedings of the Neural Information Processing—20th International Conference, ICONIP 2013, Part II, pp. 257–264. Daegu, Korea, 3–7 November 2013

References

Chapter 3
Locally Adaptive Link Prediction

Abstract In this chapter, we address the shortcomings of the existing link prediction similarity measures; every similarity measure assigns the same weight to a node irrespective of the pair of nodes between which the similarity is being computed. In contrast, the weight of a node must depend based on the local neighborhood of the node pair under consideration. In this regard, we propose the Locally Adaptive (LA) similarity measure, a generic similarity measure, which adapts (assigned weight) across different local neighborhoods by assigning different weights to the same node based on the neighborhood. Further, using a smoothening parameter, we can show that the state-of-the-art similarity measures like CN, AA, and RA are specific forms of the proposed similarity measure. Experiments on benchmark datasets show an improvement of upto 6 % using the LA similarity measure when compared to the state-of-the-art link prediction similarity measures.

Keywords Local neighborhood · Power law coefficient · Maximum likelihood estimate · Bayesian estimate · Prior distribution

3.1 Introduction

In this chapter, by *Local neighborhood* between two nodes a and b, we mean the set of nodes adjacent to both a and b. Henceforth, we shall refer to the degree distributions of the nodes in the local neighborhood and the entire network as *local degree distribution* and *global degree distribution* respectively.

The major limitations of the existing similarity measures are as follows:

1. ***Identical weight is assigned to a node irrespective of the local neighborhood between any two nodes in which it is present:*** Consider two unconnected node pairs say a_1, b_1 and a_2, b_2 both of which have different local neighborhoods but include a common neighbor z. It is natural to assume that the local neighborhoods are different between the node pairs a_1, b_1 and a_2, b_2. When we compute the similarity between a_1, b_1 and a_2, b_2 the same weight is assigned to z for both the node pairs. CN assigns to z a weight of 1 for both the node pairs. AA assigns to

© The Author(s) 2016
V. Srinivas and P. Mitra, *Link Prediction in Social Networks*,
SpringerBriefs in Computer Science, DOI 10.1007/978-3-319-28922-9_3

Fig. 3.1 A small example
network

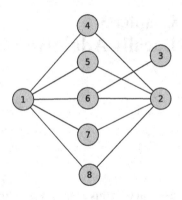

z a weight of $\frac{1}{\log(x_z)}$ for both the node pairs. Similarly, RA assigns to z a weight
of $\frac{1}{x_z}$ for both the node pairs.

Example: In Fig. 3.1, consider node pairs (1, 2) and (2, 3) and RA similarity
measure. Node 6 is a common node to both the node pairs. In computing the
similarity between nodes 1 and 2, node 6 is assigned a weight which is inversely
proportional to its degree (3). Similarly, even for computing the similarity between
nodes 1 and 3 also node 6 is assigned the same weight ($\frac{1}{3}$) based on its degree.
We observe that even in the presence of other common nodes, node 6 is assigned
a weight of $\frac{1}{3}$ while computing similarity between 1 and 2. Further, in the absence
of other common nodes between 1 and 3 also node 6 is assigned a weight of $\frac{1}{3}$.
This may not be appropriate as the importance/influence of a common node varies
across different neighborhoods. In other words, the importance of the common
node depends on its local degree in the neighborhood and also the degrees of the
end nodes between which the local neighborhood is considered. So, we will need
different weights to be assigned to the same common node when considered in
different local neighborhoods.

2. ***The performance of a similarity measure varies with the local neighborhood:***
   A similarity measure that has a good performance on a sparse local neighborhood
   may fail to do well when the local neighborhood is dense and vice versa. Thus,
   different similarity measures may be effective in different local neighborhoods;
   we need a similarity measure that can adapt.

   Example: In Fig. 3.1, we may have to compute the similarity between node pairs
   (1, 2) and (2, 3). We can observe that the size of the neighborhood between the
   node pairs is different. Between nodes 1 and 2 the size of the local neighborhood
   is high. In such cases, it is better to use weighted similarity functions like AA and
   RA to compute similarity compared to CN. However, if we consider computing
   similarity for node pair (2, 3) there is only one common node. Hence, we can
   simply use the CN similarity measure to compute similarity instead of using
   AA and RA. Thus, we observe that different similarity measures can be used
   to compute similarity in different neighborhoods. However, as already discussed
   sparsity varies across neighborhoods and networks.

3.2 Generalization of Link Prediction Similarity Measures

In order to compute the similarity between two unconnected nodes a and b, we need to find the weight assigned to each common node z. Here, by weight we refer to the amount a common node z contributes to the similarity value between a and b. Hence, we can write as follows:

$$s(a, b) = \sum_{z \in N(a) \cap N(b)} w(z) \tag{3.1}$$

where $s(a, b)$ represents the similarity between nodes a and b and $w(z)$ represents the weight assigned to the node z which is a common node. Thus, in general, we can write the weight of each common node z as follows:

$$w(z) = \beta + (1 - \beta)\frac{1}{x_z} \quad \text{for } \beta > 0 \tag{3.2}$$

This is similar to *regularization*. Overfitting is one of the main problems in building Machine Learning Models. Regularization is used to compensate for overfitting in learning models. It introduces a penalty term that penalizes for model complexity. Similarly, in computing $w(z)$, considering specifically only the degree of z can lead to overfitting. Hence, we use β to compensate for overfitting and $(1 - \beta)/\beta$ is the regularization parameter. The weight of a common node z for various similarity measures can be written as follows:

$$w(z) = 1 \text{ (CN)}$$

$$w(z) = \frac{1}{x_z} \text{ (RA)}$$

$$w(z) = \frac{1}{\log(x_z)} \text{ (AA)}$$

Comparing the above equations with Eq. 3.2, we observe the following cases:

1. $\beta = 1$: Considering $\beta = 1$ we give an equal weight to each common node ignoring the local neighborhood in which the node is present. This gives the CN similarity measure.
2. $\beta = 0$: Considering $\beta = 0$ we assign to each common node a weight which is inversely proportional to its degree. This gives the RA similarity measure.
3. $(1 - \beta) > \beta \implies 0 < \beta < 0.5$: we get the AA similarity measure. By choosing an appropriate β in this range to approximate the logarithm term in the denominator of the AA similarity measure.

Figure 3.2 shows the variation of weight assigned to a common node for various similarity measures. Note that PA measure assigns a weight of 0 to the common node; PA computes similarity between two nodes as the product of the degree of

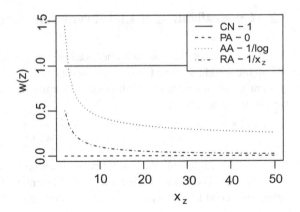

Fig. 3.2 Weight of a common node z using various similarity measures

the two nodes. Hence, we do not explicitly consider the PA similarity measure. The bottom and the top curve in the figure represent the weight assigned by the RA and CN similarity measures, respectively, to a common node; these correspond to β equal to 0 and 1, respectively. The middle curve represents the weight assigned to a common node by the AA similarity measure. From Eq. 3.2 and Fig. 3.2, we can be observe that if the weight assigned to a common node by a similarity measure is closer to the topmost curve then it implies that the similarity measure gives equal importance to all the nodes in the local neighborhood. Similarly, if the weight assigned to a common node is closer to the bottom-most curve then the similarity measure gives possibly different weights to different nodes in the local neighborhood. AA similarity measure assigns weight that falls between these two extremes.

3.3 Locally Adaptive (LA) Similarity Measure

Considering the limitations of the existing similarity measures which we discussed previously, we propose a similarity measure which we call the LOCALLY ADAPTIVE SIMILARITY MEASURE (LA). The LA similarity measure is designed based on the philosophy that the similarity measure adapts across various local neighborhoods, i.e., a node is assigned different weights across different neighborhoods. Further, we want the designed similarity measure to depend on the entire network along with the local neighborhood. The LA similarity measure is defined as follows:

$$LA(a, b) = \sum_{z \in N(a) \cap N(b)} \frac{1}{(x_z)^{c\left(\frac{\alpha_l}{\alpha_g}\right)}} \qquad (3.3)$$

where, α_l corresponds to the power law coefficient of the local degree distribution, α_g corresponds to the power law coefficient of the global degree distribution and c is the smoothening parameter ($0 \le c \le \frac{\alpha_g}{\alpha_l}$). The proposed similarity measure depends on three parameters. The impact of the parameters and their importance is discussed below:

1. α_g: Given a network the global degree distribution is known and hence α_g is fixed.
2. α_l: It varies across different local neighborhoods. Also, $\frac{\alpha_l}{\alpha_g}$ characterizes the rate at which the local degree distribution is mimicking the global degree distribution. Further, $\frac{\alpha_l}{\alpha_g}(> 0)$ varies with the local neighborhood. It permits the similarity function to adapt locally.
3. c: It is a tunable parameter which we call the smoothening parameter. It can be exploited to realize a broad range of similarity measures including and beyond CN, AA and RA.

3.3.1 Properties of the LA Similarity Measure

Some of the interesting properties of the LA similarity measure are:

1. Weight ($w(z)$) assigned to a common node z by the LA similarity measure is $\frac{1}{(x_z)^{c\left(\frac{\alpha_l}{\alpha_g}\right)}}$.
2. Parameter estimation techniques to estimate the value of α_l and α_g are discussed in Sect. 3.5. The maximum likelihood estimate of α which is represented by $\hat{\alpha}$ can be written as follows:

$$\hat{\alpha} = 1 + \frac{k}{\sum_{i=1}^{k} \ln \frac{x_i}{x_{min}}} \tag{3.4}$$

where, x is the random variable corresponding to the degree of a node in the network and k represents the size of the local neighborhood. From the above it is observed that $\hat{\alpha} > 0$. Hence, $\alpha_l/\alpha_g > 0$.

3. LA IS A GENERIC VERSION OF THE EXISTING SIMILARITY MEASURES which can be realized as follows:

   a. $c = 0$ gives CN as a specific case of LA. Note that without using the tunable parameter c it is not possible to realize CN from LA because $\frac{\alpha_l}{\alpha_g} > 0$.
   b. $c = \frac{\alpha_g}{\alpha_l}$ gives RA as a specific case of LA. Here also, without the presence of c, RA can be realized from LA only if $\alpha_l = \alpha_g$ for every neighborhood which is not realistic to assume.
   c. $0 < c < \left(\frac{\alpha_g}{\alpha_l}\right)$ gives AA similarity measure.

   For simplicity, we set $c = 1$. We leave tuning the c parameter as a part of the future work.

4. THE WEIGHT ASSIGNED TO A COMMON NODE USING LA SIMILARITY MEA- SURE VARIES ACROSS DIFFERENT LOCAL NEIGHBORHOODS. The value of α_l varies according to the local neighborhood while α_g remains constant throughout

the network. Hence, the locally adaptive similar measure adapts according to the local neighborhood.

5. Let us represent the average degree of the local and the global neighborhood by avg_l and avg_g respectively. Thus, we can simplify the above equation as

$$\frac{\alpha_l}{\alpha_g} \propto \frac{\frac{\sum_{i=1}^{k} \ln(x_i)}{n}}{\sum_{j=1}^{n} \ln(x_j)} = \frac{\frac{1}{E[\ln(x_i)]}}{\frac{1}{E[\ln(x_j)]}} \approx \frac{\frac{1}{\ln(E[x_i])}}{\frac{1}{\ln(E[x_j])}} = \frac{\frac{1}{\ln(avg_l)}}{\frac{1}{\ln(avg_g)}}$$

$$\frac{\alpha_l}{\alpha_g} \propto \frac{\ln(avg_g)}{\ln(avg_l)}$$

Note that for simpler analysis, we assume that $\ln(E[x]) \approx E[\ln(x)]$. Further, avg_g is fixed given the network. Thus, for the average degree of the local neighborhood there are three possibilities:

a. $avg_g > avg_l \implies \frac{\alpha_l}{\alpha_g} > 1$
b. $avg_g = avg_l \implies \frac{\alpha_l}{\alpha_g} = 1$
c. $avg_g < avg_l \implies \frac{\alpha_l}{\alpha_g} < 1$

Hence, we observe that $\alpha_l/\alpha_g < 1$ if the average degree of the local neighborhood is greater than the average degree of the network. Similarly, $\alpha_l/\alpha_g > 1$ if the average degree of the local neighborhood is less than the average degree of the network. The value is unity only when both the neighborhoods have the same average degree.

6. α_g is greater than α_l with high probability for scale-free networks. Having higher value of α_g compared to α_l will ensure that the LA similarity function when plotted in a similar fashion as compared to Fig. 3.2 the curve lies between CN and RA. Hence, it means that LA similarity measure will assign a weight $(w(z))$ less than 1 for a common node z.

As explained in the previous chapter, MIDT shows the best performance compared to the existing similarity measures in terms of AUC. The reason for this is that the proposed similarity measure (MIDT) assigns higher weight to the common nodes when compared to the existing similarity measures (CN, AA and RA). Thus, it desirable to have the exponent value of the degree term in the similarity measure between 0 and 1. We show that the exponent of the degree term of the proposed LA similarity measure $(c\frac{\alpha_l}{\alpha_g})$ lies between 0 and 1. We do not specifically consider c in the theorem as $0 \leq c \leq 1$. Hence, if $0 < \frac{\alpha_l}{\alpha_g} < 1 \implies 0 < c\frac{\alpha_l}{\alpha_g} < 1$.

Theorem 3.1 α_l is greater than α_g with a very small probability for scale-free networks.

Proof Using the estimate for α_l and α_g from Eq. 3.4 and using Markov Inequality, we can write as follows:

$$p(\alpha_l \geq \alpha_g) \leq \frac{E[\alpha_l]}{\alpha_g} = \frac{E\left[1 + \frac{k}{\sum_{i=1}^{k} \ln(x_i)}\right]}{\left(1 + \frac{n}{\sum_{j=1}^{n} \ln(x_j)}\right)}$$

where, k refers to the size of the local neighborhood, i.e., the number of nodes in the local neighborhood and n refers to the number of nodes in the network. We could use Markov Inequality as α_l is random and α_g is constant given the network.

As the nodes of the graph are drawn from the same distribution and are independent of one another, we can write $E[\ln(x)] = E[\ln(x_i)]$ $\forall i,\ 1 \leq i \leq k$, where x is the random variable corresponding to the degree of a node in the graph. Thus, we can simplify the above equation as,

$$p(\alpha_l \geq \alpha_g) \leq \frac{E\left[1 + \frac{k}{k \times \ln(x)}\right]}{\left(1 + \frac{n}{\sum_{j=1}^{n} \ln(x_j)}\right)} = \frac{E\left[1 + \frac{1}{\ln(x)}\right]}{\left(1 + \frac{n}{\sum_{j=1}^{n} \ln(x_j)}\right)}$$

Note that we can bound $\ln(x)$ as,

$$\frac{1}{\ln(x)} < \frac{1}{x^{\frac{1}{e^2}}}$$

Hence, we write as,

$$p(\alpha_l \geq \alpha_g) < \frac{E\left[1 + \frac{1}{x^{\frac{1}{e^2}}}\right]}{\left(1 + \frac{n}{\sum_{j=1}^{n} \ln(x_j)}\right)} = \frac{1 + E\left[\frac{1}{x^{\frac{1}{e^2}}}\right]}{\left(1 + \frac{n}{\sum_{j=1}^{n} \ln(x_j)}\right)}$$

Scale-free network follow power law degree distribution [4]. The density function of a power law degree distribution can be represented as

$$f(x) = cx^{-\alpha}$$

We can compute c by using $c \int_{x=1}^{x=max} f(x)\, dx = 1$ where *max* represents the maximum degree of any node in the network. Hence,

$$c = \frac{(1 - \alpha)}{(max^{1-\alpha} - 1)}$$

Further,

$$E\left[\frac{1}{x^{\frac{1}{e^2}}}\right] = c \int_{x=1}^{x=max} x^{-\alpha} x^{-\frac{1}{e^2}} \, dx = \frac{(1 - \alpha)}{\left(1 - \alpha - \frac{1}{e^2}\right)} \times \frac{\left(max^{\left(1-\alpha-\frac{1}{e^2}\right)} - 1\right)}{\left(max^{(1-\alpha)} - 1\right)}$$

$$p(\alpha_l \geq \alpha_g) < \frac{\left(1 + \frac{(1-\alpha)}{\left(1-\alpha-\frac{1}{e^2}\right)} \times \frac{\left(max^{\left(1-\alpha-\frac{1}{e^2}\right)} - 1\right)}{\left(max^{(1-\alpha)}-1\right)}\right)}{\left(1 + \frac{n}{\sum_{j=1}^{n} \ln(x_j)}\right)}$$

$$p(\alpha_l \geq \alpha_g) < \frac{\left(1 + \frac{(1-\alpha)}{\left(1-\alpha-\frac{1}{e^2}\right)} \times \frac{\left(max^{\left(1-\alpha-\frac{1}{e^2}\right)} - 1\right)}{\left(max^{(1-\alpha)}-1\right)}\right)}{\left(1 + \frac{1}{\ln(x)}\right)} \tag{3.5}$$

Equation 3.5 shows a bound for $p(\alpha_l \geq \alpha_g)$. The bound can be shown to be very small. Hence, we can conclude that the value of α_l is smaller than α_g with high probability.   □

3.4   Locally Adaptive Algorithm

Here, we present the outline of the link prediction algorithm using LA similarity measure. While using LA similarity measure, we will have to compute α_l and α_g. We estimate α_g using the maximum likelihood estimate (Eq. 3.4) as the number of nodes in the graph is typically high. α_l varies with the local neighborhood of nodes a and b. Hence, for estimating α_l, the maximum likelihood estimate value for α_l for the local neighborhood may not be accurate as the local neighborhood may be small. Hence, we also use Bayesian estimation technique to obtain the Maximum A Posteriori (MAP) estimate by making use of different prior distributions to estimate α_l. We proceed to estimating the power law coefficient α in the next section.

Algorithm 3: Outline of Link Prediction Algorithm using LA
---

    input   : Input Graph $G_t = (V, E_t)$
                 lp is the number of likely edges we want to predict
    output: Predicted Edges Based on Similarity
1 **for** $(a, b) \notin E_t$ **do**
2   Compute α_l and α_g which represent the power law coefficients for local and global
       degree distribution respectively using Eq. 3.4.
3   Compute the similarity score for node pair (a, b) as shown:

$$LA(a, b) = \sum_{z \in N(a) \cap N(b)} \frac{1}{(x_z)^{c\left(\frac{\alpha_l}{\alpha_g}\right)}}$$

4 **end**
5 Sort the node pairs in descending order based on the computed LA score.
6 Output the set of the top lp links.

3.5 Power Law Coefficient Estimation

From Eq. 3.3, we can observe that in order to compute the similarity using the LA similarity measure, we will have to compute α_l and α_g. We represent both α_l and α_g using a variable α and show how to estimate α. Using the same formulae, one can specifically estimate α_l and α_g. The degree distribution of a social network is typically a power law degree distribution. We can write the power law degree distribution [1] as,

$$p(x|\alpha) = \frac{\alpha - 1}{x_{min}} \times \left(\frac{x}{x_{min}}\right)^{-\alpha} \quad \text{for } x > x_{min}$$

where, α is the power law coefficient for a given x_{min} which we can consider as 1. This means that the network follows a power law distribution for $x > x_{min}$. The random variable x^1 in the above equation, represents the degree of a node and the degree distribution describes the possibility of choosing an x degree node from the network. We must note that we need to estimate the value of the power law coefficient α for the proposed function. We estimate the parameter α using the standard techniques namely maximum likelihood and Bayesian estimation, respectively, in the next two sections.

3.5.1 Maximum Likelihood Estimate of α

In this section, we proceed to estimate the power law coefficient α. The power law distribution from [1] which is a conditional density function can be expressed as follows:

---

[1]x is a random variable corresponding to the degree of a node.

$$p(x|\alpha) = \frac{\alpha - 1}{x_{min}} \times \left(\frac{x}{x_{min}}\right)^{-\alpha} \quad \text{for } x > x_{min}$$

If we have k i.i.d samples which represent k common neighbors between a pair of nodes, then we can write for the likelihood given α, α, $p(\mathcal{D}|\alpha)$ as

$$p(\mathcal{D}|\alpha) = \prod_{i=1}^{k} p(x_i|\alpha) = \prod_{i=1}^{k} \frac{\alpha - 1}{x_{min}} \times \left(\frac{x_i}{x_{min}}\right)^{-\alpha}$$

where, x_i represents the degree of the ith common neighbor and $\mathcal{D}$ represents the set of the common neighbors. Hence,

$$p(\mathcal{D}|\alpha) = \left(\frac{\alpha - 1}{x_{min}}\right)^{k} \times \prod_{i=1}^{k} \left(\frac{x_i}{x_{min}}\right)^{-\alpha}$$

Instead of maximizing the likelihood function, we can also maximize the log-likelihood function for simplicity as ln function is monotonic. Hence, we can write the log-likelihood function $\mathcal{L}(\alpha)$ as follows:

$$\mathcal{L}(\alpha) = \ln p(\mathcal{D}|\alpha)$$

$$= k \times \ln(\alpha - 1) - k \times \ln(x_{min}) - \alpha \times \sum_{i=1}^{k} \ln\left(\frac{x_i}{x_{min}}\right)$$

The maximum likelihood estimate $\hat{\alpha}$ can be written as,

$$\hat{\alpha} = \arg\max_{\alpha} \mathcal{L}(\alpha)$$

Thus, we find $\hat{\alpha}$ by taking $\frac{\partial \mathcal{L}(\alpha)}{\partial \alpha} = 0$. Hence,

$$\frac{\partial \mathcal{L}(\alpha)}{\partial \alpha} = \frac{k}{\alpha - 1} - \sum_{i=1}^{k} \ln\left(\frac{x_i}{x_{min}}\right) = 0$$

$$\implies \hat{\alpha} = 1 + \frac{k}{\sum_{i=1}^{k} \ln \frac{x_i}{x_{min}}} \tag{3.6}$$

The difficulty with the maximum likelihood estimate for α is that in the presence of a very small number of common neighbors (small k), the estimate for α will not be accurate. Hence, we will need to make use of the Bayesian technique to estimate α using various prior probability functions which we consider next.

3.5.2 Bayesian Estimate of α

We use Bayesian estimation technique to estimate α which we represent by $\hat{\alpha}_k$ using different prior probability functions $p(\alpha)$. Note that $\hat{\alpha}_k$ varies with the number of common neighbors, k. As k tends to larger values, $\hat{\alpha}_k$ will tend to the maximum likelihood estimate $\hat{\alpha}$ [2]. We use some of the standard probability density functions for the prior distribution of α.

3.5.2.1 Uniform Prior Distribution

Initially, we use the uniform prior distribution for estimating α assuming that α is equally likely to be any value between α_1 and α_2. Here, $p(\alpha)$ is defined as

$$p(\alpha) = \begin{cases} \frac{1}{\alpha_2 - \alpha_1} & \text{if } \alpha_1 \leq \alpha \leq \alpha_2 \\ 0 & \text{if } \alpha < \alpha_1 \text{ and } \alpha > \alpha_2 \end{cases}$$

for some known α_1. Using Bayes theorem, we can write as follows:

$$p(\alpha|\mathcal{D}) = k_1 \times p(\mathcal{D}|\alpha) \times p(\alpha) = k_1 \times p(\alpha) \times \prod_{i=1}^{k} p(x_i|\alpha)$$

where, k_1 is a constant for the denominator term $p(\mathcal{D})$ in the Bayes theorem. Hence,

$$p(\alpha|\mathcal{D}) = k_1 \times \frac{1}{\alpha_2 - \alpha_1} \times \prod_{i=1}^{k} p(x_i|\alpha) = k_2 \times \prod_{i=1}^{k} p(x_i|\alpha)$$

$$= k_2 \times \prod_{i=1}^{k} \frac{\alpha - 1}{x_{min}} \times \left(\frac{x_i}{x_{min}}\right)^{-\alpha}$$

The MAP estimate $\hat{\alpha}_k$ [2] can be written as

$$\hat{\alpha}_k = \arg \max_{\alpha} \ p(\alpha|\mathcal{D}) = \arg \max_{\alpha} \ \ln p(\alpha|\mathcal{D})$$

Thus, we find $\hat{\alpha}_k$ by taking $\frac{\partial \ln p(\alpha|\mathcal{D})}{\partial \alpha} = 0$. Hence, we can simplify $\ln p(\alpha|\mathcal{D})$ as follows,

$$\ln p(\alpha|\mathcal{D}) = \ln(k_2) + k \ln(\alpha - 1) - k \ln(x_{min}) - \alpha \sum_{i=1}^{k} \ln\left(\frac{x_i}{x_{min}}\right)$$

Hence,

$$\frac{\partial \ln p(\alpha|\mathcal{D})}{\partial \alpha} = \frac{k}{\alpha - 1} - \sum_{i=1}^{k} \ln\left(\frac{x_i}{x_{min}}\right) = 0$$

$$\Longrightarrow \hat{\alpha}_k = 1 + \frac{k}{\sum_{i=1}^{k} \ln \frac{x_i}{x_{min}}} \tag{3.7}$$

We observe that $\hat{\alpha}_k$ is same as $\hat{\alpha}$ for all values of k.

3.5.2.2   Power Law Prior Distribution

Power law prior distribution with power law coefficient value of α_1 is used next to estimate the value of α. Here, $p(\alpha)$ is defined as

$$p(\alpha) = (\alpha_1 - 1) \times \alpha^{-\alpha_1}$$

for some known α_1. Using Bayes theorem, we can write as follows:

$$p(\alpha|\mathcal{D}) = k_1 \times p(\mathcal{D}|\alpha) \times p(\alpha) = k_1 \times p(\alpha) \times \prod_{i=1}^{k} p(x_i|\alpha)$$

where k_1 is a constant for the denominator term $p(\mathcal{D})$ in the Bayes theorem. Hence,

$$p(\alpha|\mathcal{D}) = k_1 \times (\alpha_1 - 1) \times \alpha^{-\alpha_1} \times \prod_{i=1}^{k} p(x_i|\alpha)$$

$$= k_2 \times \alpha^{-\alpha_1} \times \prod_{i=1}^{k} p(x_i|\alpha)$$

for some constant α_1. The MAP estimate $\hat{\alpha}_k$ can be written as

$$\hat{\alpha}_k = \arg\max_{\alpha} \ p(\alpha|\mathcal{D}) = \arg\max_{\alpha} \ \ln p(\alpha|\mathcal{D})$$

Thus, we find $\hat{\alpha}_k$ by taking $\frac{\partial \ln p(\alpha|\mathcal{D})}{\partial \alpha} = 0$. We can simplify $\ln p(\alpha|\mathcal{D})$ as follows:

$$\ln p(\alpha|\mathcal{D}) = \ln(k_2) - \alpha_1 \ln(\alpha) + k \ln(\alpha - 1)$$

$$- k \ln(x_{min}) - \alpha \sum_{i=1}^{k} \ln\left(\frac{x_i}{x_{min}}\right)$$

Hence,

$$\frac{\partial \ln p(\alpha|\mathcal{D})}{\partial \alpha} = -\frac{\alpha_1}{\alpha} + \frac{k}{\alpha - 1} - \sum_{i=1}^{k} \ln\left(\frac{x_i}{x_{min}}\right) = 0$$

$$\implies \hat{\alpha}_k = \frac{z_1 + \sqrt{z_1^2 + 4\alpha_1 \sum_{i=1}^{k} \ln\left(\frac{x_i}{x_{min}}\right)}}{2 \sum_{i=1}^{k} \ln\left(\frac{x_i}{x_{min}}\right)} \tag{3.8}$$

where $z_1 = (k + \sum_{i=1}^{k} \ln\left(\frac{x_i}{x_{min}}\right) - \alpha_1)$. For large values of k, we can write Eq. 3.8 as

$$\hat{\alpha}_k = \frac{1}{2}(z_2 + 1) + \frac{1}{2}\sqrt{(z_2 + 1)^2} = 1 + z_2$$

where $z_2 = \frac{k}{\sum_{i=1}^{k} \ln\left(\frac{x_i}{x_{min}}\right)}$.

$$\implies \hat{\alpha}_k = 1 + \frac{k}{\sum_{i=1}^{k} \ln\left(\frac{x_i}{x_{min}}\right)}$$

We observe that $\hat{\alpha}_k$ is same as $\hat{\alpha}$ for large values of k.

3.5.2.3 Exponential Prior Distribution

Next, we use an exponential distribution as the prior distribution to estimate α. α_1 is used as a prior mean for the exponential distribution. Here, $p(\alpha)$ is defined as

$$p(\alpha) = \begin{cases} \alpha_1 e^{-\alpha\alpha_1} & \text{if } \alpha \geq 0 \\ 0 & \text{if } \alpha < 0 \end{cases}$$

for some known α_1. Using Bayes theorem, we can write as follows:

$$p(\alpha|\mathcal{D}) = k_1 \times p(\mathcal{D}|\alpha) \times p(\alpha) = k_1 \times p(\alpha) \times \prod_{i=1}^{k} p(x_i|\alpha)$$

where, k_1 is a constant for the denominator term $p(\mathcal{D})$ in the Bayes theorem. Hence,

$$p(\alpha|\mathcal{D}) = k_1 \times \alpha_1 \times e^{-\alpha\alpha_1} \times \prod_{i=1}^{k} p(x_i|\alpha)$$

$$= k_2 \times e^{-\alpha\alpha_1} \times \prod_{i=1}^{k} p(x_i|\alpha)$$

for some constant α_1. The MAP estimate $\hat{\alpha}_k$ can be written as

$$\hat{\alpha}_k = \arg \max_{\alpha} \; p(\alpha|\mathcal{D}) = \arg \max_{\alpha} \; \ln \, p(\alpha|\mathcal{D})$$

Thus, we find $\hat{\alpha}_k$ by taking $\frac{\partial \ln p(\alpha|\mathcal{D})}{\partial \alpha} = 0$. We can simplify $\ln \, p(\alpha|\mathcal{D})$ as follows,

$$\ln p(\alpha|\mathcal{D}) = \ln(k_2) - \alpha\alpha_1 + k \times \ln(\alpha - 1)$$
$$-k \times \ln(x_{min}) - \alpha \times \sum_{i=1}^{k} \ln\left(\frac{x_i}{x_{min}}\right)$$

Hence,

$$\frac{\partial \ln p(\alpha|\mathcal{D})}{\partial \alpha} = -\alpha_1 + \frac{k}{\alpha - 1} - \sum_{i=1}^{k} \ln\left(\frac{x_i}{x_{min}}\right) = 0$$

$$\implies \hat{\alpha}_k = 1 + \frac{k}{\alpha_1 + \sum_{i=1}^{k} \ln \frac{x_i}{x_{min}}} \tag{3.9}$$

We observe that $\hat{\alpha}_k$ is same as $\hat{\alpha}$ for large values of k.

3.5.2.4 Normal Prior Distribution

We use a normal distribution as the prior distribution to estimate α with a prior mean α_1 and standard deviation σ_1. Here, $p(\alpha)$ is defined as,

$$p(\alpha) = \frac{1}{\sigma_1\sqrt{2\pi}} \times e^{-\frac{1}{2}\left(\frac{\alpha-\alpha_1}{\sigma_1}\right)^2}$$

for some known α_1 and σ_1. Using Bayes theorem, we can write as follows:

$$p(\alpha|\mathcal{D}) = k_1 \times p(\mathcal{D}|\alpha) \times p(\alpha) = k_1 \times p(\alpha) \times \prod_{i=1}^{k} p(x_i|\alpha)$$

where, k_1 is a constant for the denominator term $p(\mathcal{D})$ in the Bayes theorem. Hence,

$$p(\alpha|\mathcal{D}) = k_1 \times \frac{1}{\sigma_1\sqrt{2\pi}} \times e^{-\frac{1}{2}\left(\frac{\alpha-\alpha_1}{\sigma_1}\right)^2} \times \prod_{i=1}^{k} p(x_i|\alpha)$$
$$= k_2 \times e^{-\frac{1}{2}\left(\frac{\alpha-\alpha_1}{\sigma_1}\right)^2} \times \prod_{i=1}^{k} p(x_i|\alpha)$$

$$= k_2 \times e^{-\frac{1}{2}\left(\frac{\alpha-\alpha_1}{\sigma_1}\right)^2} \times \prod_{i=1}^{k} \frac{\alpha-1}{x_{min}} \times \left(\frac{x_i}{x_{min}}\right)^{-\alpha}$$

$$= k_2 \times e^{-\frac{1}{2}\left(\frac{\alpha-\alpha_1}{\sigma_1}\right)^2} \times \left(\frac{\alpha-1}{x_{min}}\right)^k \times \prod_{i=1}^{k} \left(\frac{x_i}{x_{min}}\right)^{-\alpha}$$

The Bayesian estimate $\hat{\alpha}_k$ can be written as

$$\hat{\alpha}_k = \arg\max_{\alpha}\ p(\alpha|\mathcal{D}) = \arg\max_{\alpha}\ \ln\ p(\alpha|\mathcal{D})$$

Thus, we find $\hat{\alpha}_k$ by taking $\frac{\partial\ \ln\ p(\alpha|\mathcal{D})}{\partial\ \alpha} = 0$. We can simplify $\ln\ p(\alpha|\mathcal{D})$ as follows,

$$\ln\ p(\alpha|\mathcal{D}) = \ln(k_2) - \frac{1}{2}\left(\frac{\alpha-\alpha_1}{\sigma_1}\right)^2 + k \times \ln(\alpha-1)$$
$$- k \times \ln(x_{min}) - \alpha \times \sum_{i=1}^{k} \ln\left(\frac{x_i}{x_{min}}\right)$$

Hence,

$$\frac{\partial\ \ln\ p(\alpha|\mathcal{D})}{\partial\ \alpha} = -\frac{\alpha-\alpha_1}{\sigma_1^2} + \frac{k}{\alpha-1} - \sum_{i=1}^{k}\ln\left(\frac{x_i}{x_{min}}\right) = 0$$

Hence, if we substitute for $\sigma_1^2 \times \sum_{i=1}^{k} \ln\left(\frac{x_i}{x_{min}}\right)$ by some constant k_3, we get a quadratic in α as follows,

$$\alpha^2 + (k_3 - \alpha_1 - 1)\alpha + (\alpha_1 - k_3 - k\sigma_1^2) = 0$$

$$\implies \hat{\alpha}_k = \frac{-z_3 + \sqrt{z_3^2 - 4(\alpha_1 - k_3 - k\sigma_1^2)}}{2} \tag{3.10}$$

where, $z_3 = (k_3 - 1 - \alpha_1)$. As k tends to larger values, we can write the above equation as,

$$\hat{\alpha}_k = \frac{-k_3 + \sqrt{k_3^2 + 2(1-\alpha_1)k_3 + 4k\sigma_1}}{2}$$

Dividing throughout by k_3 gives,

$$\hat{\alpha}_k = -\frac{1}{2} + \frac{\sqrt{k_3^2 + 2(1 - \alpha_1)k_3 + 4k\sigma_1^2}}{2k_3}$$

$$\hat{\alpha}_k = -\frac{1}{2} + \frac{1}{2}\sqrt{1 + 4\frac{k}{\sum_{i=1}^{k} \ln\left(\frac{x_i}{x_{min}}\right)^2}}$$

From the above discussion, we showed how α parameter is estimated using various parameter estimation techniques which is used in the LA similarity measure. Let us call the LA similarity measure which uses the Maximum Likelihood Estimate for α_l by LA^{MLE}. Similarly, let us call the LA similarity measure which uses the MAP Estimate (Exponential Distribution Prior) for α_l by LA^E, (Power Law Distribution Prior) for α_l by LA^P and (Normal Distribution Prior) for α_l by LA^N. We ignore the LA similarity measure which uses the uniform prior probability distribution to estimate α (Eq. 3.7) as the estimated value is same as the maximum likelihood estimate of α (Eq. 3.6).

Note that α_l value when estimated using uniform, exponential and power law prior distributions converges to the α_l estimated using MLE when the number of common neighbors is large.

3.6   Experimental Setup

We conduct the experiments in the same way as described in Sect. 2.4. We discuss the results in the next section.

3.7   Results

In this section, we present results on using LA similarity measure against the existing popular similarity measures in terms of accuracy. Precisely, we use maximum likelihood and MAP estimate values for α in the LA similarity measures and present results using different estimation techniques and prior distributions explicitly.

From Tables 3.1 and 3.2, we observe that LA-based measures perform better than PA, CN, AA, and RA. We show the best results on each dataset when performing 50–50 and 80–20 edge sampling in boldface. Note that G_t is more dense in the case of 80–20 edge sampling when compared to 50–50 edge sampling. Hence, we can note that the link prediction similarity measures show a better performance of AUC in the case of 80–20 edge sampling when compared to 50–50 edge sampling.

Table 3.1 AUC results for 50–50 edge sampling

Predictor	Amazon	CondMat	HepTh
PA	12.1	36.84	48.78
CN	30.3	64.9	59.13
AA	69.89	73.52	69.65
RA	69.13	68.45	65.65
LA_{MLE}	**71.89**	**75.27**	71.53
LA_P	71.59	74.64	70.13
LA_E	71.85	73.48	69.88
LA_N	71.79	75.01	**71.93**

Table 3.2 AUC results for 80–20 edge sampling

Predictor	Amazon	CondMat	HepTh
PA	8.99	54.36	41.82
CN	29.48	63.86	57.41
AA	46.61	78.78	73.74
RA	43.85	81.53	72.18
LA_{MLE}	48.15	**84.54**	**79.2**
LA_P	**48.2**	83.86	74.75
LA_E	47.8	84.06	74.96
LA_N	47.59	83.61	73.57

We observe that the proposed LA similarity measure performs the best among all the similarity measures. We observe that in most cases, LA similarity measure using MLE estimate performs the best. The reason for this could be that the local neighborhoods are dense on average, ensuring that α_l estimated using MLE is good. Note that the AUC performance has increased by up to 6 % using LA-based similarity measures.

3.8 Summary

In this chapter, we first discussed the shortcomings of the existing local similarity measures in link prediction context and proposed a locally adaptive similarity measure (LA). It addresses many of the limitations associated with the existing similarity measures. The proposed similarity measure locally adapts by weighing the contribution of the nodes in computing similarity based on the neighborhood considered. We further show that it is a generalized form of the existing similarity measures using the smoothening parameter. Specifically, we observed an improvement of upto 6 % in terms of AUC on the benchmark datasets. Large scale of experimentation shows that the LA similarity measure has an effective and consistent performance in terms of AUC. In the future, we would like to examine the role of the smoothening parameter to improve the performance of LA-based measures.

Similarity measures perform poorly on sparse graphs, which have a small CC [3] and the performance increases as the CC of the graph increases and sparseness of the graph reduces. In the next chapter, we present a two-phase approach which uses an auxiliary graph to predict links instead of graph G_t to deal with the challenges posed by sparse networks.

References

1. Clauset, A., Shalizi, C., Newman, M.E.J.: Power-law distributions in empirical data. SIAM Rev **51**(4), 661–703 (2009)
2. Duda, R.O., Hart, P.E., Stork, D.G.: Pattern Classification. Wiley, New York (2012)
3. Feng, X., Zhao, J., Xu, K.: Link prediction in complex networks: a clustering perspective. Eur. Phys. J. B **85**(1), 3 (2012)
4. Newman, M.E.J.: Networks: An Introduction. Oxford University Press Inc, New York (2010)

Chapter 4
Two-Phase Framework for Link Prediction

Abstract In this chapter, we focus only on link prediction in sparse networks. Link prediction in sparse networks poses a major challenge as link prediction similarity measures perform poorly on sparse networks (Eur. Phys. J. B, 85(1):3, 2012, [2]). Similar work has been done earlier in link prediction in (Proceedings of the 16th ACM SIGKDD International Conference on Knowledge Discovery and Data Mining, pp. 393–402, 2010, [3]) which they term as "cold start link prediction problem." The author considers predicting the network structure using available information regarding the nodes when the whole network is missing. In a similar manner, we predict the whole network when the network is evolving and very sparse. Specifically, we propose a two-phase framework to predict links in sparse networks. The generality of our approach makes it feasible to use it along with any link prediction similarity measures. Experiments on benchmark datasets show the superiority of the framework as it shows an improvement of upto 47 % in terms of AUC.

Keywords Two-phase link prediction · Clustering coefficient · Kullback–Leibler (KL) divergence · Optimization problem · Boost graph · Sparse network

4.1 Introduction

Social networks are typically sparse in nature. Clustering coefficient is a good indicator of the density of the network. The connectivity structure of a sparse graph does not provide sufficient local neighborhood information. Hence, using the similarity function between nodes in a sparse network may not give satisfactory results. It is also known that similarity measures perform poorly on graphs having a small CC [2].

In this chapter, we propose an approach for predicting links by paying attention to CC. We do not consider the zero-degree nodes; we preprocess the network data to remove such nodes. According to [4], in principle a new link is created when it is likely to form a clique in a given network or which forms as many cliques as possible in the network. From this we can conclude that new links are added in such a way that the CC of the network increases. Increase in CC corresponds to an increase

---

Material in this chapter appeared in [6].

© The Author(s) 2016

V. Srinivas and P. Mitra, *Link Prediction in Social Networks*,
SpringerBriefs in Computer Science, DOI 10.1007/978-3-319-28922-9_4

in the number of cliques or near-cliques in the network. Increase in the number of cliques models in an appropriate manner the network evolution process. Hence, preprocessing the network to increase its CC might be the right way to proceed in link prediction on sparse networks.

Motivated by this idea, in our work, we add some important nonexistent links to the sparse graph G_t which gives us a new graph G_{t^*} which we call as **Boost Graph**; G_{t^*} has more relevant information pertaining to local neighborhood and higher clustering coefficient compared to G_t due to addition of relevant new links. We use the connectivity structure of G_{t^*} in computing the similarity between the node pairs for link prediction.

4.2 Motivation

Consider the following notation for understanding the section better:

Notation

- $G_t = (V, E_t)$—network at the current time
- $G_{t'} = (V, E_{t'})$—network at a future time $(t' > t)$
- $G_{t^*} = (V, E_{t^*})$—boost network for $(t' > t^* > t)$
- $KL1$—KL divergence of G_t $(KL(G_t))$
- α—power law coefficient of graph $G_{t'}$
- β—power law coefficient of graph G_{t^*}
- $p(x)$—degree distribution of $G_{t'}$; $p(x) \propto x^{-\alpha}$
- $q(x)$—degree distribution of G; $q(x) \propto x^{-\beta}$

In the link prediction problem, we need to predict the links that are likely to get added to network G_t at a later point in time, i.e., we need to predict the links present in $G_{t'}$ but not in G_t. In general, the link prediction algorithm uses similarity measures (CN, AA etc.) to detect the missing links using the structure of graph G_t. Instead, it would be better to add most likely missing links to G_t which are likely to form in the future. This gives our boost graph G_{t^*}; G_{t^*} will have a smaller KL value compared to G_t and larger than that of $G_{t'}$. Similarly, G_{t^*} will have higher CC value compared to G_t and smaller than that of $G_{t'}$. Thus, we can use G_{t^*} to predict the links of graph $G_{t'}$ more effectively as G_{t^*} is better than G_t in terms of CC and KL values for link prediction. Link prediction similarity measures show good performance when the CC of the graph is higher. We show the existence of such a G_{t^*} theoretically, which satisfies the properties discussed above.

4.2.1 Theoretical Existence of G_{t^*}

As discussed above, we need to find a G_t^* which satisfies the following constraints:

$$CC(G_t) < CC(G_{t^*}) < CC(G_{t'})$$
$$KL(G_t) > KL(G_{t^*}) > KL(G_{t'})$$

In this section, we show that there exists an optimal G_{t*} which when used for link prediction can be better than G_t. We formulate it as

$$G_{t*} = \underset{G}{\operatorname{argmax}} \quad CC(G)$$

There arises a difficulty with the problem formulation. The reason for this is that the maximum value of CC(G) is obtained when we reach the complete graph. However, for a complete graph, the KL divergence value will be very high and link prediction to form a complete graph is not useful. Hence, we have to optimize in such a way that CC of G_{t*} increases and KL of G_{t*} reduces compared to G_t.

Hence, this **optimization problem** can be formulated as follows:

$$G_{t*} = \underset{G}{\operatorname{argmax}} \quad CC(G)$$
$$\text{subject to} \quad KL(G) < KL1$$

We can rewrite the above problem as

$$G_{t*} = \underset{G}{\operatorname{argmax}} \quad CC(G)$$
$$\text{subject to} \quad KL(G) + \varepsilon \leq KL1$$

Solution: The Lagrangian for the above problem can be written as

$$L(\beta, \lambda) = CC(G) + \lambda(KL1 - \varepsilon - KL(G)) \tag{4.1}$$

From Eq. 4.1, we can observe that λ is the balancing factor which controls the rate of increase of CC value and rate of decrease of KL Divergence. So, we need to find the optimal λ.

We consider the lower bound as 2 for the degree henceforth as we consider only nodes having degree of at least 2 in our graph. Now, for sparse graphs, the local clustering coefficient correlates negatively with the degree [1], i.e., in specific it varies as k^{-1} for a k degree node. Thus, we can write the CC value as follows:

$$CC(G) = \int_{k=2}^{k=\infty} \frac{q(k)}{k} \, dk$$

Scale-free social networks which are usually sparse follow the power law degree distribution. Thus, $q(k)$ follows a power law degree distribution using which we can approximately write the above equation as

$$CC(G) = \int_{k=2}^{k=\infty} \frac{k^{-\beta}}{k} \, dk = \left. \frac{k^{-\beta}}{-\beta} \right|_2^{\infty} = \frac{1}{\beta 2^\beta} \tag{4.2}$$

as $max \gg 2$. Similarly, we can solve for KL divergence of G with respect to $G_{t'}$ as follows:

$$KL(G) = \int_{k=2}^{k=\infty} p(k) \log \frac{p(k)}{q(k)} \, dk$$

Using the power law degree distribution for graphs G and $G_{t'}$, we can rewrite the above equation as

$$KL(G) = \int_{k=2}^{k=\infty} k^{-\alpha} \log \frac{k^{-\alpha}}{k^{-\beta}} \, dk = (\beta - \alpha) \int_{k=2}^{k=\infty} k^{-\alpha} \log(k) \, dk$$

$$= (\beta - \alpha) \left[-\frac{k^{1-\alpha}((\alpha - 1) \log k + 1)}{(\alpha - 1)^2} \Bigg|_2^\infty \right]$$

We can simplify the above equation as follows:

$$KL(G) = \frac{(\beta - \alpha)\alpha}{(\alpha - 1)^2 \, 2^{\alpha-1}} = (\beta - \alpha)C$$

$$\text{where, } C = \frac{\alpha}{(\alpha - 1)^2 \, 2^{\alpha-1}} \tag{4.3}$$

From Eqs. 4.1, 4.2 and 4.3 we get,

$$L(\beta, \lambda) = \frac{1}{\beta \, 2^\beta} + \lambda(KL1 - \varepsilon - (\beta - \alpha)C) \tag{4.4}$$

Partially differentiating L with respect to λ we get

$$\frac{\partial L}{\partial \lambda} = (KL1 - \varepsilon - (\beta - \alpha)C) = 0$$

$$\implies (\beta - \alpha)C = KL1 - \varepsilon$$

We can simplify for optimal β (β^*) as

$$\beta^* = \alpha + \frac{(KL1 - \varepsilon)}{C} \tag{4.5}$$

Similarly partially differentiating L with respect to β, we get

$$\frac{\partial L}{\partial \beta} = \frac{\partial}{\partial \beta} \left(\frac{1}{\beta 2^\beta} \right) + \lambda C = 0$$

$$\implies -\frac{\log(2)}{\beta 2^\beta} - \frac{1}{\beta^2 \, 2^\beta} + \lambda C = 0$$

We can simplify the above equation as

$$-\frac{(1+\beta)}{\beta^2\,2^\beta} + \lambda C = 0$$

$$\implies \lambda C = \frac{(1+\beta)}{\beta^2\,2^\beta}$$

Thus, we can write λ using optimal value of β as

$$\lambda^* = \frac{(1+\beta^*)}{C\beta^{*2}\,2^{\beta^*}} \tag{4.6}$$

From Eqs. 4.5 and 4.6, we can write λ^* in terms of α and KL1 only and thus we can find λ^* so that we can maximize CC(G) and minimize KL(G); and thus end up with optimal G_t^* having degree distribution coefficient β^*. This shows that there exists a G_t^* which satisfies the constraints.

4.3 Two-Phase Framework

Note that link prediction similarity measures use the information between node pairs for similarity computation. Further, in the case of sparse networks or evolving networks at an early stage the local neighborhood may not give enough information. Thus, link prediction similarity measures may have a poor performance.

We propose a two-phase link prediction framework for sparse networks. Let *lp-factor* denote a value between 0 and 1 which represents the fraction of missing links we want to add to graph G_t to construct the boost graph (G_t^*). Let S be a similarity measure which takes two nodes x and y as input and returns the similarity between x and y. The outline of the two-phase framework is presented as follows: From the point of notation, note that in the two-phase framework a similarity measure $S_1\_S_2$ means that S_1 similarity measure is applied in phase1 and S_2 similarity measure is applied in phase2. Further, observe that instead of using the structure of graph G_t, we use graph G_{t^*} for link prediction; we compute the similarity for all unconnected node pairs in G_t even though they are connected in G_{t^*}.

4.4 Experimental Setup

For conducting our experiments, we used the datasets shown in Table 1.1. The datasets do not contain time stamps representing the time at which links are formed in the network. For evaluating the link prediction similarity measures on such graph datasets, we employ the experimental setup reported in [5]. We conduct experiments on sampled training and test graphs generated in the following ways:

Algorithm 4: Outline of Two-Phase Framework

---

   input : Input Graph $G_t = (V, E_t)$

           Boost Graph $G_{t^*} = (V, E_{t^*})$

           Similarity measure $S_1\_S_2$

           lp is the number of likely edges we want to predict

           lp-factor is the fraction of lp links that we add to G_t to generate G_{t^*}

   output: Predicted Edges Based on Similarity

1 **phase1 starts here.**
2
3 **for** $(a, b) \notin E_t$ **do**
4   compute similarity as $S_1(a, b)$ using the common nodes information from graph G_t.
5 **end**
6 Sort the node pairs in descending order based on the computed score.
7 Add the top lp-factor $\times$ lp number of links to E_t to form E_t^*, where $0 \leq$ lp-factor ≤ 1.
8
9 **phase1 ends here.**
10
11 **phase2 starts here.**
12
13 **for** $(a, b) \notin E_t$ **do**
14   compute similarity as $S_2(a, b)$ using the common node information from graph G_{t^*}
15 **end**
16 **phase2 ends here.**
17
18 Output the top lp links.

---

1. Perform edge sampling to divide the dataset two parts each having 10 and 90 % links. We use the smaller part as the training graph (G_t) and the larger part as the test graph $(G_{t'})$. We predict the edges of $G_{t'}$. Let us call this **10–90 edge sampling**. 10–90 edge sampling indicates that training graph G_t has 10 % links and test graph $G_{t'}$ has the remaining 90 % links. This simulates link prediction on sparse networks.

2. Perform edge sampling to divide the dataset two parts each having 20 and 80 % links. We use the smaller part as the training graph (G_t) and the larger part as the test graph $(G_{t'})$. We predict the edges of $G_{t'}$. Let us call this **20–80 edge sampling**. 20–80 edge sampling indicates that training graph G_t has 20 % links and test graph $G_{t'}$ has the remaining 80 % links. This simulates link prediction on sparse networks.

We use G_t as the graph at the current time instance and predict the links of the test graph $(G_{t'})$ to validate the predictions. We repeat this process 10 times to reduce any statistical bias introduced due to sampling.

Link prediction similarity measures are evaluated using the AUC metric as explained in Sect. 2.4. We discuss the results in the next section.

Table 4.1 AUC results for 10–90 edge sampling

Predictor	Amazon	CondMat	HepTh
PA	45.68	52.8	51.8
CN	9.47	16.9	19.81
AA	4.31	38.18	14.3
RA	4.42	37.57	14.19
PA_PA	44.82	52.82	51.17
PA_CN	4.45	9.68	12.64
PA_AA	*55.66*	*69.66*	*52.69*
PA_RA	*55.73*	*69.56*	*51.78*
CN_PA	52.81	43.97	44.38
CN_CN	12.21	18.98	20.73
CN_AA	*56.02*	*70.83*	*51.79*
CN_RA	**74.39**	**72.83**	**62.88**
AA_PA	43.26	51.57	49.24
AA_CN	4.47	8.51	13.82
AA_AA	4.39	38.53	13.42
AA_RA	4.16	37.58	12.79
RA_PA	47.32	50.51	48.12
RA_CN	4.22	7.63	12.53
RA_AA	4.36	38.02	14.37
RA_RA	4.51	37.34	14.38

4.5 Results

On performing the experiments using various similarity measures in the two-phase
framework, we report the AUC results in Tables 4.1 and 4.2 on various datasets on
predicting 90 and 80 % missing links, respectively.

From Tables 4.1 and 4.2, we observe that PA performs better than CN, AA, and
RA in the case of sparse networks. Note that for the experiments we set lp-factor
to 0.5. Further, employing the two-phase framework using the existing similarity
measures shows significant improvement when compared to the existing link prediction similarity measures. We show the best results on each dataset when performing
10–90 and 20–80 edge sampling in boldface. We also show in italic font the results
which are better than the existing similarity measures in the tables. Note that G_t
is more dense in the case of 20–80 edge sampling when compared to 10–90 edge
sampling. Hence, we can note that the link prediction similarity measures show a
better performance of AUC in the case of 20–80 edge sampling when compared to
10–90 edge sampling.

We can observe that using PA in phase 1 shows improvement in performance when
followed up with AA or RA in phase 2. However, best results were attained when

Table 4.2 AUC results for 20–80 edge sampling

Predictor	Amazon	CondMat	HepTh
PA	24.55	47.51	52.13
CN	34.47	47.97	47.78
AA	16.25	55.48	46.02
RA	15.75	53.05	42.13
PA_PA	24.72	47.7	51.26
PA_CN	13.86	26.32	31.48
PA_AA	*59.74*	*75.93*	*67.07*
PA_RA	*59.91*	*74.97*	*65.24*
CN_PA	46.18	63.72	64.06
CN_CN	33.75	48.21	49.15
CN_AA	**81.03**	*80.07*	*69.15*
CN_RA	45.59	**81.43**	**73.95**
AA_PA	45.59	54.45	58.4
AA_CN	15.9	25.72	31.17
AA_AA	16.25	55.51	46.24
AA_RA	15.64	53.06	38.69
RA_PA	51.59	55.07	53.96
RA_CN	15.94	54.99	22.42
RA_AA	15.84	54.99	44.51
RA_RA	16.16	53.22	42.64

we use CN in phase1 and follow it up with AA or RA in phase2. This shows that CN and PA show good performance in sparse networks and AA and RA show good performance in denser networks. This is consistent with the results shown in earlier chapters. Further, two-phase framework enhances the link prediction performance both in the cases of 10–90 and 20–80 edge sampling. Further, two-phase framework shows better performance in the case of 20–80 edge sampling when compared to 10–90 edge sampling. In other words, two-phase framework shows better performance when the training graph is dense. Note that the AUC performance has increased by up to 47 %. Next, we show how the link prediction performance varies with lp-factor.

4.5.1 Variation of Performance with lp-Factor

We observe from our results that lp-factor is an important parameter in the proposed approach. We show the results by tabulating AUC versus lp-factor on various datasets on predicting 90 and 80 % missing links in Tables 4.3, 4.4 and 4.5. In general, we observe that the AUC varies by varying lp-factor uniformly across all the datasets. Although, the variation in AUC is not very significant, in general, we observe that

Table 4.3 AUC versus lp-factor for Amazon dataset

lp-factor	10–90 edge sampling					20–80 edge sampling				
	0.2	0.4	0.6	0.8	1.0	0.2	0.4	0.6	0.8	1.0
PA_PA	46.54	46.81	46.79	47.21	46.44	25.95	25.29	25.31	24.16	24.61
PA_CN	4.03	4.38	4.05	4.2	4.12	14.05	13.76	13.99	14.38	14.32
PA_AA	56.12	56.88	56.62	56.64	55.69	59.03	58.89	59.66	59.53	60.21
PA_RA	55.69	56.74	56.76	55.13	56.93	59.76	59.21	58.88	59.6	58.54
CN_PA	46.33	46.25	46.78	46.94	47.36	33.53	33.94	34.76	34.26	33.69
CN_CN	5.32	5.82	5.29	5.39	4.95	16.53	17.02	16.19	17.02	17.41
CN_AA	56.69	56.13	56.57	55.99	56.38	55.88	56.12	55.5	55.82	55.94
CN_RA	56.02	56.78	56.71	56.19	57.02	56.98	56.24	57.07	57.14	57.61
AA_PA	47.82	48.45	48.04	48.13	47.05	48.79	50.03	50.25	49.53	50.17
AA_CN	3.64	4.03	3.64	3.88	3.71	15.57	15.37	15.94	15.44	15.86
AA_AA	4.02	3.69	4.03	3.6	4.19	16.86	15.47	16.62	16.02	15.81
AA_RA	3.71	3.82	3.77	3.59	3.7	15.52	15.28	15.1	15.87	15.54
RA_PA	48.12	48.37	47.87	47.8	47.91	50.18	50.09	49.92	49.99	50.63
RA_CN	3.92	3.73	3.56	3.57	4.08	15.2	15.84	15.35	15.42	16.18
RA_AA	4.12	3.66	3.8	3.51	3.63	16.07	16.13	16.03	16.06	16.04
RA_RA	3.74	3.97	3.83	3.65	3.99	15.65	15.76	15.42	15.49	15.91

Table 4.4 AUC versus lp-factor for CondMat dataset

lp-factor	10–90 edge sampling					20–80 edge sampling				
	0.2	0.4	0.6	0.8	1.0	0.2	0.4	0.6	0.8	1.0
PA_PA	53.29	53.65	53.7	54.27	53.75	47.59	46.29	48.19	47.55	47.05
PA_CN	9.59	9.43	9.74	9.81	9.97	25.59	26.23	26.11	25.95	25.83
PA_AA	70.12	70.27	70.64	69.75	70.36	73.78	74.32	74.8	74.28	74.65
PA_RA	70.14	69.7	69.59	69.41	69.05	74.07	74.03	73.67	74.4	74.12
CN_PA	49.74	49.83	50.38	50.28	50.36	38.59	37.24	37.85	37.5	39.29
CN_CN	15.42	15.5	14.74	15.33	14.68	45.45	46.02	44.67	44.77	45.78
CN_AA	71.18	71.0	71.78	70.18	71.22	79.08	79.31	79.78	78.62	79.29
CN_RA	71.13	71.76	70.38	70.76	71.03	79.61	79.25	79.16	79.15	79.39
AA_PA	51.95	51.73	51.15	51.7	51.57	55.05	54.62	54.8	55.4	54.51
AA_CN	9.01	8.91	8.61	8.47	9.03	56.68	57.72	57.85	56.42	57.57
AA_AA	37.38	37.56	37.01	36.44	36.73	26.54	26.03	26.08	26.03	26.04
AA_RA	36.34	36.29	36.1	35.83	35.75	55.18	54.31	54.61	54.75	55.18
RA_PA	50.89	51.95	51.21	51.13	51.81	56.16	54.4	54.97	55.68	54.97
RA_CN	7.72	7.73	7.87	7.27	7.95	24.88	24.95	25.92	25.3	24.92
RA_AA	36.45	36.51	36.65	36.94	35.32	56.64	56.05	57.57	56.63	56.42
RA_RA	35.86	36.31	36.23	35.86	36.1	55.18	54.31	54.63	54.18	54.25

Table 4.5 AUC versus lp-factor for HepTh dataset

lp-factor	10–90 edge sampling					20–80 edge sampling				
	0.2	0.4	0.6	0.8	1.0	0.2	0.4	0.6	0.8	1.0
PA_PA	49.8	51.42	48.79	50.54	49.58	48.19	48.03	47.55	47.47	47.89
PA_CN	7.01	7.76	7.24	7.43	7.09	32.98	32.38	32.32	32.53	32.64
PA_AA	51.83	52.58	53.11	53.53	51.51	68.39	69.34	68.3	68.9	68.36
PA_RA	51.94	51.77	51.85	52.38	52.17	66.46	66.02	67.48	66.3	66.37
CN_PA	51.26	52.55	50.58	50.76	50.94	44.82	44.12	45.22	44.39	44.45
CN_CN	8.16	8.16	8.68	9.09	9.12	40.5	40.26	40.48	40.64	41.36
CN_AA	54.44	50.76	53.19	53.6	53.91	70.07	70.17	70.83	69.67	70.42
CN_RA	50.53	50.67	51.29	51.84	50.68	67.66	67.98	67.38	67.81	68.37
AA_PA	48.16	47.31	46.83	47.78	47.96	57.39	57.66	56.91	58.27	57.21
AA_CN	8.2	7.89	8.03	7.8	8.03	32.36	32.88	32.12	32.2	32.19
AA_AA	13.73	13.79	13.83	13.57	13.67	45.86	46.03	45.19	44.73	46.24
AA_RA	14.28	14.48	14.28	14.04	14.09	37.35	37.88	37.86	37.86	38.33
RA_PA	46.83	47.73	48.22	47.78	48.73	53.74	53.5	53.22	53.5	53.58
RA_CN	8.31	8.1	8.56	8.71	8.71	23.3	23.42	23.49	24.38	23.8
RA_AA	14.11	13.84	14.17	14.11	14.17	42.97	43.01	43.78	43.51	43.43
RA_RA	13.63	13.78	14.12	14.19	14.84	40.18	41.79	40.72	41.22	41.32

AUC increases with increasing the value of lp-factor. However, in some cases, AUC increases till certain point and falls as lp-factor gets increased. Hence, on a safer note, we set the value of lp-factor to 0.5 for experimentation. *The results indicate that PA and CN are more effective on sparse graphs while AA and RA are effective on denser graphs.*

4.6 Summary

In this chapter, we presented a two-phase framework which shows significant improvement compared to the standard link prediction approach on sparse networks. Specifically, in our approach we add missing links which are more likely to occur in the future to the existing sparse network and make it denser to form the boost graph. We exploit the connectivity structure of this denser network for effective link prediction. The exhaustive experimentation using our approach demonstrates the superiority and robustness of our approach. Specifically, we show a significant improvement of upto 47 % on benchmark datasets. In the future, we would like to test for the efficacy of this method in terms of increase in the number of phases of the two-phase algorithm to form a multiphase link prediction. Further, we would like to work on how to determine the value of lp-factor based on the network at hand as opposed to setting empirically.

References

1. Bloznelis, M.: Degree and clustering coefficient in sparse random intersection graphs. Ann. Appl. Prob. **23**(3), 1254–1289 (2013)
2. Feng, X., Zhao, J., Xu, K.: Link prediction in complex networks: a clustering perspective. Eur. Phys. J. B **85**(1), 3 (2012)
3. Leroy, V., Cambazoglu, B.B., Bonchi, F.: Cold start link prediction. In: Proceedings of the 16th ACM SIGKDD International Conference on Knowledge Discovery and Data Mining, pp. 393–402. ACM (2010)
4. Liu, Z., He, J., Srivastava, J.: Cliques in complex networks reveal link formation and community evolution. ArXiv preprint arXiv:1301.0803 (2013)
5. Lü, L., Zhou, T.: Link prediction in complex networks: a survey. Phys. A **390**(6), 1150–1170 (2011)
6. Virinchi, S., Mitra, P.: Two-phase approach to link prediction. In: Proceedings of the 21st International Conference on Neural Information Processing, ICONIP 2014, Part II, pp. 413–420. Kuching, Malaysia, 3–6 November 2014

Chapter 5
Applications of Link Prediction

Abstract Link prediction has a wide variety of applications. Graphs provide a natural abstraction to represent interactions between different entities in a network. We can have graphs representing social networks, transportation networks, disease networks, email/telephone calls network to list a few. Link prediction can specifically be applied on these networks to analyze and solve interesting problems like predicting outbreak of a disease, controlling privacy in networks, detecting spam emails, suggesting alternative routes for possible navigation based on the current traffic patterns, etc.

Keywords Recommender system · Spam mail detection · Citation network · Influence detection · Disease prediction

In this chapter, we discuss the applications of link prediction in practical cases.

5.1 Recommender Systems

Recommender systems are information filtering systems that recommend new products to the users based on the users' previous rating or preference to similar products. Although, recommender systems have been approached using classical collaborative filtering algorithms [21], link prediction has been successfully applied to recommender systems to generate quality recommendations.

Link prediction has been widely applied in *recommender systems* [6, 11]. Even though collaborative filtering algorithms have been widely applied in the context of recommender systems, they are greatly limited by the sparse data problem. Link prediction proximity measures outperformed the standard collaborative filtering algorithms when used for providing recommendations. Li and Chen [14] propose using a graph kernel-based recommendation framework to predict possible user-item interactions. This framework demonstrates improved performance when a large number of recommendations are required. Li and Chen [15] map transactions to a bipartite user-item interaction graph, thereby converting the recommendation problem to a link prediction problem. Li et al. [16] calculate domain similarities between products to weigh the product recommendations to customers with larger weights whose categories are more similar to the users' preferences to improve recommendation quality. Further work in this direction can be found in [2, 5, 12, 20, 22, 24, 26].

5.2   Spam Mail Detection

It is common to receive spam mails to ones mail account. These mails are unwanted mails which are not normal and unexpected for the users soliciting regular mails. In this context, link prediction has been applied to *detect anomalous mails* for traffic monitoring purpose over various communication channels in [10] using graph-theoretic approaches. They specifically model the one-to-many relationship between a sender and multiple recipients.

5.3   Privacy Control in Social Networks

It is important to preserve privacy in any network by shielding users from unreliable users. For example, for administrators of social networks like Facebook, it would be important to hide important user information like email, phone number, photos, etc. from unreliable users as it can manipulated. Link weight in social networks indicates the level of trust between two users. Al-Oufi et al. [1] propose a capacity-based algorithm which adopts Advogato trust metric for identifying people of trust based on weighted relationships. Thus, for a given user, the proposed method finds all possible trustworthy users who are connected to the given user in the network, thereby shielding the given user from unreliable users. This ensures privacy control for the users in social networks.

5.4   Identifying Missing References in a Publication

A publication might contain link to others work, in such a case, others work must be cited to acknowledge their contribution. However, it is possible that a publication can have missing references. In such a case, it is important to identify missing references in a publication to avoid plagiarism. In this context, Kc et al. [13] provide a framework for the generation of links between referenced and otherwise interlinked documents. This framework proposes using the nodes of the graph for documents, and links in the graph for the references between the documents. Using this graph, it is possible to obtain a set of possible references for a new document. It can be applied in predicting missing or useful references for a new document.

5.5 Expert Detection

This problem deals with identifying expert(s) in a given domain. For example, given any network which corresponds to links between various researchers, we need to find the experts from the network in various domains. In this context, link prediction has been applied in co-authorship networks for *finding domain experts* in [19]. Liu and Ning [17] show how link prediction can be used to rank candidates for high-level government posts.

5.6 Influence Detection

Given a social network, it is important to understand which users are most influential in the network. For example, in case of sales of a phone, influential users in a network can have a significant impact on the sales of the product. The impact of a highly influential user in favor of the phone can incur more sales of the phone in the network. In contrast, the impact of a highly influential user who is against the phone can reduce the sales of the phone in the network. Cervantes et al. [3] introduce local measures to estimate the influence of users in a collaborative network. Their approach adds and removes each node iteratively from the network and link prediction is simultaneously performed to understand the influence of the node in the network. Every collaborator is represented by a vertex. Similarly, Nguyen et al. [18] estimate a person's influence and personality traits using link prediction.

5.7 Routing in Networks

The problem deals with identifying optimal routes in networks to improve routing performance. Frequent breaks in routes in mobile ad hoc networks adversely affect the quality of mobile wireless networks thereby posing a challenge. Weiss et al. [23] and Yadav et al. [25] propose methods to estimate signal strength-based link availability prediction for routing. Estimate of link breakage time based on link information allows local route repair or new route discovery for the packets, thereby reducing packet drops and end-to-end delay. Hu and Hou [9] propose traffic prediction approach using link prediction improving better routing of packets in wireless networks. Further work in this direction can be found in [4, 8].

5.8 Disease Prediction

Folino and Pizzuti [7] work on applying link prediction models to predict the onset
of diseases given the current health status of the patients. They propose to construct a
comorbidity network where nodes are the diseases and an edge between two diseases
represents the simultaneous occurrence of the two diseases in a patient. Standard link
prediction techniques are applied on the network to output a ranked list of scores
between two diseases. The higher the link score, the higher is the likelihood of the
co-occurrence of the two diseases. This technique can reveal morbidities a patient
could develop in the future.

References

1. Al-Oufi, S., Kim, H., El-Saddik, A.: Controlling privacy with trust-aware link prediction in
   online social networks. In: ICIMCS 2011, The Third International Conference on Internet
   Multimedia Computing and Service, pp. 86–89. Chengdu, China, 5–7 August 2011
2. Benchettara, N., Kanawati, R., Rouveirol, C.: A supervised machine learning link prediction approach for academic collaboration recommendation. In: Proceedings of the 2010 ACM
   Conference on Recommender Systems, RecSys 2010, pp. 253–256. Barcelona, Spain, 26–30
   September 2010
3. Cervantes, E.P., Mena-Chalco, J.P., Oliveira, M.C.F., Cesar, R.M.: Using link prediction to
   estimate the collaborative influence of researchers. In: 9th IEEE International Conference on
   eScience, eScience 2013, pp. 293–300. Beijing, China, 22–25 October 2013
4. Chen, J., Han, Y., Li, D., Nie, J.: Link prediction and route selection based on channel state
   detection in uasns. IJDSN **2011** (2011)
5. Chiluka, N., Andrade, N., Pouwelse, J.A.: A link prediction approach to recommendations
   in large-scale user-generated content systems. In: Advances in Information Retrieval—33rd
   European Conference on IR Research, ECIR 2011, Proceedings, pp. 189–200. Dublin, Ireland,
   18–21 April 2011
6. Esslimani, I., Brun, A., Boyer, A.: Densifying a behavioral recommender system by social
   networks link prediction methods. Social Netw. Analys. Mining **1**(3), 159–172 (2011)
7. Folino, F., Pizzuti, C.: Link prediction approaches for disease networks. In: Information Technology in Bio- and Medical Informatics - Third International Conference, ITBAM 2012, Proceedings, pp. 99–108. Vienna, Austria, 4–5 September 2012
8. Han, Q., Bai, Y., Gong, L., Wu, W.: Link availability prediction-based reliable routing for
   mobile ad hoc networks. IET Commun. **5**(16), 2291–2300 (2011)
9. Hu, C., Hou, J.C.: A link-indexed statistical traffic prediction approach to improving IEEE
   802.11 PSM. Ad Hoc Netw. **3**(5), 529–545 (2005)
10. Huang, Z., Zeng, D.D.: A link prediction approach to anomalous email detection. In: Proceed-
    ings of the IEEE International Conference on Systems, Man and Cybernetics, pp. 1131–1136.
    Taipei, Taiwan, 8–11 October 2006
11. Huang, Z., Li, X., Chen, H.: Link prediction approach to collaborative filtering. In: ACM/IEEE
    Joint Conference on Digital Libraries, JCDL 2005, Proceedings, pp. 141–142. Denver, CO,
    USA, 7–11 June 2005
12. Huang, K., Fan, Y., Tan, W., Li, X.: Service recommendation in an evolving ecosystem: a
    link prediction approach. In: 2013 IEEE 20th International Conference on Web Services, pp.
    507–514. Santa Clara, CA, USA, 28 June–3 July 2013

13. Kc, M., Chau, R., Hagenbuchner, M., Tsoi, A.C., Lee, V.C.S.: A machine learning approach to link prediction for interlinked documents. In: Focused Retrieval and Evaluation, 8th International Workshop of the Initiative for the Evaluation of XML Retrieval, INEX 2009, Revised and Selected Papers, pp. 342–354. Brisbane, Australia, 7–9 December 2009

14. Li, X., Chen, H.: Recommendation as link prediction: a graph kernel-based machine learning approach. In: Proceedings of the 2009 Joint International Conference on Digital Libraries, JCDL 2009, pp. 213–216. Austin, TX, USA, 15–19 June 2009

15. Li, X., Chen, H.: Recommendation as link prediction in bipartite graphs: a graph kernel-based machine learning approach. Decis. Support Syst. 54(2), 880–890 (2013)

16. Li, J., Zhang, L., Meng, F., Li, F.: Recommendation algorithm based on link prediction and domain knowledge in retail transactions. In: Proceedings of the Second International Conference on Information Technology and Quantitative Management, ITQM 2014, National Research University Higher School of Economics (HSE), pp. 875–881. Moscow, Russia, 3–5 June 2014

17. Liu, J., Ning, K.: Applying link prediction to ranking candidates for high-level government post. In: International Conference on Advances in Social Networks Analysis and Mining, ASONAM 2011, pp. 145–152. Kaohsiung, Taiwan, 25–27 July 2011

18. Nguyen, T., Phung, D.Q., Adams, B., Venkatesh, S.: Towards discovery of influence and personality traits through social link prediction. In: Proceedings of the Fifth International Conference on Weblogs and Social Media, Barcelona, Catalonia, Spain, 17–21 July 2011

19. Pavlov, M., Ichise, R.: Finding experts by link prediction in co-authorship networks. In: Proceedings of the 2nd International ISWC+ASWC Workshop on Finding Experts on the Web with Semantics, pp. 42–55. Busan, Korea, 12 November 2007

20. Pujari, M., Kanawati, R.: A supervised machine learning link prediction approach for tag recommendation. In: Online Communities and Social Computing - 4th International Conference, OCSC 2011, Held as Part of HCI International 2011, Proceedings, pp. 336–344. Orlando, FL, USA, 9–14 July 2011

21. Sarwar, B., Karypis, G., Konstan, J., Riedl, J.: Item-based collaborative filtering recommendation algorithms. In: Proceedings of the 10th international conference on World Wide Web, pp. 285–295. ACM (2001)

22. Wang, X., He, D., Chen, D., Xu, J.: Clustering-based collaborative filtering for link prediction. In: Proceedings of the Twenty-Ninth AAAI Conference on Artificial Intelligence, pp. 332–338. Austin, Texas, USA, 25–30 January 2015

23. Weiss, E., Kurowski, K., Hischke, S., Xu, B.: Avoiding route breakage in ad hoc networks using link prediction. In: Proceedings of the Eighth IEEE Symposium on Computers and Communications (ISCC 2003), pp. 57–62. Kiris-Kemer, Turkey, 30 June–3 July 2003

24. Wu, S., Raschid, L., Rand, W.: Future link prediction in the blogosphere for recommendation. In: Proceedings of the Fifth International Conference on Weblogs and Social Media, Barcelona, Catalonia, Spain, 17–21 July 2011

25. Yadav, A., Singh, Y.N., Singh, R.R.: Improving routing performance in AODV with link prediction in mobile adhoc networks. Wirel. Pers. Commun. 83(1), 603–618 (2015)

26. Yang, X., Zhang, Z., Wang, K.: Scalable collaborative filtering using incremental update and local link prediction. In: 21st ACM International Conference on Information and Knowledge Management, CIKM'12, pp. 2371–2374. Maui, HI, USA, 29 October–02 November 2012

Chapter 6
Conclusion

Abstract Link prediction problem has been studied extensively in relation to various applications. However, the role of degree distribution of the networks have not been explicitly exploited in this context. In this book, we deal with link prediction by considering the power law degree distribution of large-scale networks. We summarize our key findings and also discuss possible future directions in the context of link prediction.

Keywords Adamic adar index · Resource allocation index · Hybrid scheme · Outlier node · Influential node

In this book, we have examined the link prediction problem using similarity measures which compute closeness of node pairs; two nodes which are similar tend to connect to each other. We extensively worked on real-world networks. These networks follow the power-law degree distribution.

We proposed three algorithms that characterize similarity using the proposed similarity measures in link prediction for dense networks: *degree-thresholding-based similarity measures (MIDT), common edge-based measures (CNC, AAC, RAC), and locally adaptive similarity measure (LA)*. Further, we present a two-phase-based framework for link prediction in sparse networks.

We summarize the key findings from our work as follows:

- AA and RA work well among the existing similarity measures in the case of dense networks.
- Link prediction in sparse networks is difficult when compared to link prediction in dense networks; neighborhood information in sparse networks may not be adequate compared to that in dense networks. The other reason could be that link prediction similarity measures perform poorly on networks having low CC when compared to networks having high CC.
- In general, all the proposed algorithms perform better than the popular baseline similarity measures in terms of AUC.
- LA similarity measure is the most generic similarity measure which adapts locally by taking into consideration the local degree distribution of node pairs. It can be viewed as a generic form of the existing similarity measures.

© The Author(s) 2016
V. Srinivas and P. Mitra, *Link Prediction in Social Networks*,
SpringerBriefs in Computer Science, DOI 10.1007/978-3-319-28922-9_6

- The performance of MIDT similarity measures shows that the low-degree common nodes should be given more weight/importance; high-degree common nodes can be assigned a lower weight or completely ignored in computing similarity between unconnected node pairs.
- Increasing the CC of the sparse networks will allow us to capture the similarity better, thereby, improving the performance of the link prediction similarity measures. This is the motivation for exploring a two-phase link prediction.
- Among the proposed algorithms, the LA similarity performs the best in terms of accuracy in dense networks.

Future Work

It would be good to explore the following in the context of link prediction:

- A hybrid scheme that makes use of two-phase link prediction algorithm and the LA similarity measure.
- Can we use the proposed similarity measures to detect communities, classification, and identifying influential nodes in networks? This question is meaningful as similarity forms the basis for the above-mentioned tasks.
- To devise link prediction schemes in networks where the strength/weight associated with a link varies over time. Specifically, new links could appear in the future, and the existing links may become weak or vanish in the future.
- Identifying outlier nodes in a network, where any link from any node to an outlier node is termed as a bad link. Removal of bad links in a network can be analogous to noise reduction in the classification context.
- Link prediction techniques for multiplex and multilayer networks.

Glossary

α	Power law coefficient of network.
α_l	Power law coefficient of network of local neighborhood.
α_g	Power law coefficient of network of global neighborhood.
E_t	Edges of G_t.
$E_{t'}$	Edges of $E_{t'}$.
G_t	Network at time t.
$G_{t'}$	Network at time t' ($t' > t$).
G_{t^*}	Boost/Auxiliary graph after adding lp-factor edges to G_t.
lp	Number of edges needed to predict.
$lp\text{-}factor$	Fraction between 0 and 1 for two-phase framework.
$N(a)$	Neighbors of node a.
T	Degree threshold for MIDT and Clique based approach determined using Markov Inequality.
x_a	Degree of node a.

© The Author(s) 2016
V. Srinivas and P. Mitra, *Link Prediction in Social Networks*,
SpringerBriefs in Computer Science, DOI 10.1007/978-3-319-28922-9

Index

© The Author(s) 2016
V. Srinivas and P. Mitra, *Link Prediction in Social Networks*,
SpringerBriefs in Computer Science, DOI 10.1007/978-3-319-28922-9

Printed in the United States
By Bookmasters